Praise for The Tarball Chronicles

"For those interested in putting the Gulf crisis in perspective, there can be no better guide than this funny, often uncertain, frank, opinionated, always curious, informed and awestruck accounting of how we've gone wrong and could go right—a full-strength antidote to the Kryptonite of corporate greed and human ignorance."

—*Atlanta Journal-Constitution*

"In this highly readable account, David Gessner considers the *Deepwater Horizon* oil spill as a symptom of even bigger economic and cultural challenges that loom in our future. This excellent book is not judgmental, but thought provoking and well worth reading."

—David Allen Sibley, author of *The Sibley Guide to Birds*

"I loved *The Tarball Chronicles*. David Gessner caught just the right tone and touch to help us explore this disaster."

—Nikki Giovanni, author of *Rosa*

"Brilliant—the best and most original writing coming out of the Gulf."

—*OnEarth* magazine

"Plenty of people are writing about the BP oil disaster, but few indeed will be able to make us feel the reality of it like David Gessner can. The likelihood that his account will also be action-filled and darkly funny is pure bonus."

—John Jeremiah Sullivan, author of *Pulphead*

"An eye-opening, jaw-dropping account. Gessner crafts a powerfully informative but also immensely relatable narrative. Entertaining and rousing."

—*Mother Nature Network*

"David Gessner is on a roll."

—*New Orleans Times-Picayune*

"Anyone who wanted a first-hand look at the Gulf after the news cycle ended will find it here. Brilliant, thoughtful."
—*Publishers Weekly* (starred review)

"If you read only one book about the *Deepwater Horizon* oil spill this year, it should be this one. If you plan not to read any books about it, make an exception for this blunt, funny, eye-opening quest to find the real stories behind the Gulf crisis."
—*Shelf Awareness*

"Expressive and adventurous. A profoundly personal inquiry into the aftermath of the *Deepwater Horizon* catastrophe unique in its hands-on immediacy and far-ranging ruminations."
—*Booklist*

"An expert naturalist, he not only observes but talks with people who are in the know—forceful, insightful, blood-and-guts people who will speak their minds (like David). There is grit and heartbreak and energy in just about everything he writes."
—Clyde Edgerton, author of *Lunch at the Piccadilly*

"Gessner has the heart and mind of an investigative journalist. Not everyone will be pleased with this Jeremiah in our midst, but the word is a fire and a hammer, and Gessner delivers it well."
—*Mobile Press-Register*

"Highly readable, strongly recommended."
—Fred Kasten, WWNO's *The Sound of Books*

"Vivid, funny, opinionated, poignant, and mold breaking—Gessner takes us deep into the environmental and personal tragedies of the spill."
—Jim Campbell, author of *The Final Frontiersman*

"Gessner's account of his journey blazes out with a fiery, pugilistic style. His journey around the Gulf of Mexico offers us a powerful and sobering reminder that whether or not we feel the direct effects of the oil spill in our backyards, we are all implicated, all compromised, and—most important for Gessner—all connected."

—*America, The National Catholic Weekly*

"*The Tarball Chronicles* is well worth your time. It's a darkly entertaining tribute to the Gulf coast, our 'national sacrifice zone.' But be warned: You'll come away discomfited and with more than a few questions of your own."

—*Tampa Bay Online*

Praise for My Green Manifesto *by David Gessner*

"A wonderfully readable book. Gessner's attempts to define the role of the new environmental warrior, both in terms of idealism and political practicality, are heartfelt and informed. [*My Green Manifesto*] is brave enough and intelligent enough to embrace technology as well as art, pure ideology as well as compromise, hope as well as despair, depression and paralysis as well as valor and joy."

—*Boston Globe*

"Raw and honest, there's a lilt in his jig that many will find invigorating."

—*Los Angeles Times*

"For nature-writing enthusiasts, Gessner needs no introduction. His books and essays have in many ways redefined what it means to write about the natural world, coaxing the genre from a staid, sometimes wonky practice to one that is lively and often raucous."

—*Washington Post*

"Funny and inspiring, Gessner believes that committing to a lifelong environmental fight is an act of personal fulfillment. *My Green Manifesto* is a pleasurable read, with an environmental message that there is still transcendence to be found in the 'limited wild' of our own communities. So get out there, enjoy it, and fight for it before it's gone because, at least according to Gessner, this is the key to a better life."

—*Publishers Weekly* (starred review)

"An engaging book with a serious message."

—*Kirkus Reviews*

"Earthy and funny, frank and pragmatic. Gessner asserts that nature is necessary for our well-being, that 'the most important wilderness is the one closest to home,' and that effective environmentalism is rooted not in theory, renunciation, or gloom, but, rather, in love and wonder, even anger."

—*Booklist*

"Gessner has chopped down the strangling beanstalk of environmentalism, and has merrily, adroitly, hungrily planted something new in its place. Gessner is not saying anybody is off the hook, but he offers a more effective way of relating to nature—no, in fact, of being nature."

—Craig Childs, author of *The Animal Dialogues*

"David Gessner re-invents the environmental manifesto for people who hate the word environmental as much as they hate the word manifesto. Make no mistake—he can write about a blue heron or an osprey with the best of them—but if you're looking for mystical rhapsodies to Mother Earth, go elsewhere. Gessner is convinced that re-connecting ourselves with nature doesn't start with finger-wagging; it starts with fun."

—Ginger Strand, author of *Inventing Niagara*

"So, again, what's so funny? Well, Gessner is. His writing is irreverent, to the point where some of what he writes can't be quoted in a family newspaper."

—*Cape Cod Times*

The Tarball Chronicles

Also by David Gessner

The Tarball Chronicles

A Journey Beyond the Oiled Pelican
and into the Heart of the Gulf Oil Spill

David Gessner

milkweed
editions

Some names in this book have been changed to protect
the professional status of those referred to or quoted.

Published 2012 by Milkweed Editions
Printed in the United States of America
Cover design and art by Tonky Designs
Interior design by Connie Kuhnz
The text of this book is set in Minion Pro.
12 13 14 15 16   5 4 3 2 1
First Paperback Edition
ISBN: 978-1-57131-337-9

Please turn to the back of this book for a list of the sustaining
funders of Milkweed Editions.

The Library of Congress has catalogued the hardcover edition as
follows:

Gessner, David, 1961–
    The tarball chronicles : a journey beyond the oiled pelican and into
the heart of the gulf oil spill / David Gessner.
       p.   cm.
    ISBN 978-1-57131-333-1 (alk. paper)
    1. BP Deepwater Horizon Explosion and Oil Spill, 2010.   2. Oil
spills—Environmental aspects—Mexico, Gulf of.   3. Oil wells—
Mexico, Gulf of—Blowouts.   4. Oil pollution of the sea—Mexico,
Gulf of.   5. Mexico, Gulf of--Environmental conditions.   I. Title.
    TD427.P4G47 2011
    363.738'20916364—dc23

                             2011024701

This book is printed on acid-free paper.

To Nina,
who lets me roam — and come back

THE TARBALL CHRONICLES

Firm ground is not available ground.

— *A. R. Ammons*

I hear comments sometimes that large oil companies are greedy companies, or don't care, but that is not the case in BP. We care about the small people.

— *Carl-Henric Svanberg, former chairman of BP*

The Tarball Chronicles

A Journey Beyond the Oiled Pelican
and into the Heart of the Gulf Oil Spill

PRELUDE: INTO THE GULF

It is June and you are at a cookout at a friend's house, a bar-beque with all the kids playing in the backyard. You have just gotten back from traveling and you are happy to be home. For the last fifty-nine days millions of gallons of oil have been gushing into the Gulf of Mexico, but that is not your concern, not your problem. You want nothing to do with yet another dismal, depressing environmental story. You live in North Carolina and the Gulf is almost a thousand miles away. Yes, you care about the environment, so you *should* be thinking about the oil spill, but you've put on blinders, as you often do when the harsh light of big news events blares down on you. There is too much to think about, after all, and right now you are looking out at your daughter jump-ing on a trampoline, and the spill is the furthest thing from your mind. You drink your second beer and think that life is pretty good, pretty good indeed.

But then suddenly a friend is standing in front of you, and he insists on talking about the spill. He tells you of a live video stream he has seen from a mile below the surface and of the sight of a single curious eel peering at black-red goo pouring from the spill's source, the busted Macondo well. He wonders what it is like for the people living down in the Gulf, and despite yourself and the beer and the sun on your face and your happy daughter playing, you start to wonder too. "You should be down there," he says. "You write about nature." You start to explain that that is not the kind of nature you write about—you write about birds and the coast, and you are not a journalist who chases stories. But then you stop explaining, and defending, and think simply: "Maybe he's right."

Over the next week the idea builds in your head. Maybe the Gulf *is* where you should be. Summer plans, family plans, rearrange themselves in your brain. You have a somewhat strained talk with your wife about your new plans, and, since there is no other way to get there on short notice, you decide to drive. "When will you go?" your wife asks, and it turns out your answer is "Right away."

A magazine gives you an assignment to cover the looming fall bird migration, but this is about more than birds, you know that already. When you finally decide to leave you do so in a mad rush, throwing everything in the back of your car and heading out without any real plan. Of course you are aware of the hypocrisy of traveling eight hundred miles in a vehicle powered by a refined version of the same substance that is still pouring out into the Gulf waters—but now you are driven. Now you *need* to see the oil. You're not sure why. You have heard the Gulf called a "national sacrifice zone," and maybe you want to explore this idea of sacrifice, of giving up some of our land, and our people, so the rest of us can keep living the way we do. So you go down, heading toward the Gulf.

Which gives you some idea of how I found myself sitting in a booth at an Applebee's on the border of South Carolina and Georgia. My waiter, a chipper young man named George, asked me where I was heading, and when I told him, in a somewhat reluctant and grumbling fashion, I expected him to chirp "Great!" and hurry off to get my fries and beer. Instead he thought for a minute before launching into a little sermon.

"We think it's happening *down there*," he said, jerking his thumb behind him, toward the wall with the sports posters on it. "But we're part of it, too."

I snapped to attention.

"It's all tied together," he said. "The oil we use in our cars and the oil that's washing up on our beaches."

I felt like standing up and clapping. How did this George understand something that the major media outlets couldn't seem to grasp? It turned out this was another reason I had decided to throw everything in the car and head south. It seemed that no one in the national media was writing the bigger story, or at least the longer story, and I was pretty sure that the one oil-covered bird they kept trotting out for TV was not the story. "No more bullshit" was my blunt, businessman father's favorite saying. No more bullshit indeed. This time around I would experience a story firsthand instead of letting the national media take me on its knee, like a kindly uncle, and tell me its sweet and homogenized version of the truth.

The next morning I drove through Georgia, mulling over my Applebee's epiphany. My waiter reminded me of another natural philosopher, John Muir, who traveled this same route by foot nearly 150 years ago. "When we try to pick out anything by itself," Muir said, "we find it hitched to everything else in the universe."[1] Yes . . . synapses snapped and connections crackled as the miles passed and I drank too much coffee. Wasn't the spill hitched to everything? Already the disaster seemed to be trying to teach me something, in dramatic fashion, a lesson that the world kept teaching me but that I had been slow to learn: on this planet nothing is apart from anything else—all of us, human, plant, animal: intertwined.

I drove all day. A friend in Mobile, Alabama, had offered me a place to sleep, and I had planned to go there, but then fate, in the form of weather, decided otherwise. A thunderstorm bullied me eastward, toward the Florida Panhandle, until I noticed a sign for a beach I had seen on the news a week before. That beach was known for its famously white sand, at least until the caramel sludge started washing over it. As I continued along the shore I noticed dozens of cleanup workers wavering through the mist like a ghostly prison crew in fluorescent vests, sweeping the sands. There were a couple hundred workers in all, and at first the scene made no sense, the people seemingly disembodied and floating. I soon learned that these workers—many formerly unemployed and mostly men—were being paid eighteen bucks an hour by BP to pick at the sand. I also learned that their job was to gather tarballs and toss them into huge plastic bags, then bring the bags to the command center where they were weighed and hauled away in trucks. Almost all of the workers were black, and I got the feeling that most of these men, many from the nearby city of Pensacola, hadn't spent a lot of time at this particularly touristy beach before the last few weeks.

The second level of command—the sergeants—didn't exactly look like beachgoers either. I pulled over and watched them for a while. Muscular but overweight—almost all white, incidentally—they barked orders and drove around in four-wheel-drive ATVs that looked like amped-up golf carts. At first I thought that they might enjoy their little taste of authority, but they never smiled. No one seemed to be having a good time.

"Don't ask them any questions," the girl back at the beer store had told me. "BP won't let them talk to civilians."

Her wording had sounded strange at first, but not after I saw the workers spread over the beach: they did, in fact, seem like a sluggish, corporate army.

I moved on to explore the campsites, scouting for a place to put up my tent, before eventually heading back to the beach. Once there I walked down to the water's edge, my first real encounter with the spill, and found the sand covered with tarballs. Though they didn't look like balls exactly. The small ones looked like dried rabbit turds or kernels of a not-particularly appetizing cereal. The larger ones were maps of rust-brown countries or jigsaw puzzle pieces, some the size of cow patties.

Other than the workers, I found only one person on the beach: a man sitting in a foldout chair and pointing a camera at the water. James was a surfer whose skin had been burned a crisp brown over the years. He was also an amateur photographer. He told me that the local surf report now specified where the oil was and wasn't each day, but that one day, when he went to one of the spots where it supposedly wasn't, his wife's white bathing suit turned gray.

We had been talking for a little while when a truck with Texas plates pulled into the parking lot and a small family— dad, mom, and daughter—piled out. Before long the little girl was swimming in the oily water, holding hands with her mom. I wondered if I should say something. Then the man, the father, called out to them, not out of concern for their safety, but because he had stumbled upon a giant tarball. James had shown me the same tarball earlier, saying it was the biggest he had yet seen and suggesting that he should bring it to the attention of the EPA officials. But the Texas guy was now laying claim to James's specimen, apparently interested in taking it as a souvenir or trophy. James was a peaceful man and simply shrugged when the Texas guy picked up the lump of oil and carted it off. The little girl, now out of the water, became defensive when she heard me suggest that James had rightful claim to the tarball.

"If he touches that tarball my daddy will kick his ass," she said.

She couldn't have been older than eight. The man placed the tarball in the truck bed, a toxic prize for home.

"They're real good on the grill with a little paprika!" James yelled after them as they drove off.

Soon after James packed it in, closing his chair and taking his tripod and saying good-bye. I continued down the beach, staring out at the green surf. While I expected to find ugliness here, what surprised me was the beauty. When I had first driven through the gates to this beach—a park designated as a national seashore and therefore undeveloped—something lifted inside of me. Wind swept across the thin scrap of land and birds carved up the air. It felt like the end of the earth, which it was. There were no buildings in sight, just one road running down between a spine of dunes, and my car splashed through puddles of saltwater. The surf frothed and sand blew from the ocean back to the sea oats and marsh that made up the island's leeward side, and terns—sharp angular birds like living check marks—rose up from the dunes and screeched defensively, protecting their colony. As someone who has always loved the ocean, I understand the appeal of crossing over a bridge or going through a gate and finding a place apart, a place to get away from the human world and into the world of birds, water, wind, and sand. Despite everything, that was how I felt upon entering the park. It was thrilling, really. But a "place apart" also implies seclusion and separateness from the world. It went without saying that, at that moment, the beach surrounding me was anything but.

Eventually I came upon a particularly ugly patch of tar, something the Texas family might have wanted to lay claim to. As I looked into these clumps of oily turds I began to suspect that this time we had really done it. I thought: We have passed a point, and the fact that many of our current actions are suicidal must be becoming obvious to even the most casual observer of the natural world. We have soiled ourselves. Less elegantly put, we have shat ourselves.

These were the kind of things bubbling in my head as I walked from the beach up the driveway to the ranger station. And then I saw it: This . . . what? . . . this symbol of what has gone awry. It was a truck, a white truck, oversized and muscular, larger than any truck has a right to be, with its motor running and windows up, air-conditioning blasting. I'd vowed not to use my AC during this trip, and had kept to that vow so far. It was just a symbolic gesture—being that I had just driven eight hundred miles down to the Gulf—but gestures felt like the only meaningful actions available. I knew I had no right to be outraged.

But outraged I was. When I inquired inside the station it turned out that the truck belonged to a fellow journalist, a guy making a documentary about the tarballs. Shouldn't he, of all people, get it? He was talking to the rangers, something about the evils of BP, and I didn't know what got into me—it's not the kind of thing I usually do—but before I could stop myself I interrupted him by saying: "So maybe you should shut off your truck when you're not in it?"

Right away I felt bad and he got defensive, muttering about how hot it had been out on the beach and how his personal energy use was "just a drop in the bucket." Soon

I found myself wishing I hadn't said anything and feeling guilty when the guy stormed off.

I expected that the rangers would be put off by my behavior, but instead they acted amused. After the truck guy was gone, one of them turned to me with an unexpected smile and started talking. Right away I liked her—she had sympathetic eyes, a quick wit, and an obvious love for the beach. Before long she was showing me pictures of what she called "her" beach, in the same way someone else might show you pictures of her kids. She held out one of a hidden place she went to every day for lunch, tucked between the dunes, a place far away from the tourists where she could "get away." But the tarballs had discovered her secret. Yesterday she'd had an apple with lunch and showed me a picture of a huge tarball dwarfing the leftover core.

As we talked she informed me that the fort that gave the seashore its name was built after the War of 1812 to protect our young republic from British invasion. Which had worked pretty well until three weeks ago.

She also told me about watching as the BP supervisors and contractors and their work crews descended on the park. It was a strange twist. "The British are coming!" she'd wanted to yell. And then, before she knew it, the British were here.

"We used to be in charge of this beach," she continued. "Now we are bossed around by these people. But they don't know the place. They trample things. At first they ran right over the turtle nests in their ATVs."

She described the dark day the caramel-colored tide came in and coated her beach.

"It was the color and texture of a Baby Ruth bar left out on a hot dashboard," she said.

What she felt was a sense of creeping powerlessness.

"It's not just the oil on the beach, it's the fact that we're not in control. We're like people in an occupied country."

We exchanged e-mail addresses and phone numbers and promised to keep in touch. Before I left, though, I asked her if I was allowed to swim in the water. I have traveled the coasts a lot in the past few years and at every beach I've visited—from Alaska to Cape Cod to Nova Scotia to the Outer Banks—I have made it a point to mark my arrival with a plunge into the ocean. I may not have liked seeing the little Texas girl playing in the surf but that didn't mean I didn't want to see myself in it.

She shrugged.

"We don't make the rules anymore, but BP's party line is: 'If it looks clean you can go in.'"

"That's pretty scientific."

"Yup," she replied, shaking her head but still smiling.

After bidding her good-bye I hiked back down through the dunes. Before long I felt better, swinging my arms and looking out at the water. The world opened up: a rush of wind and salt spiced with the vague smell of oil. An osprey flew above the surf, peering down for fish. Up the beach from the tarballs was a sign for a sea turtle nest where, below the ground, a newborn Kemp's ridley lay snug in egg and sand. I knew that soon enough the turtle would emerge and begin its crawl to the oily sea.

Far off in the distance, I could still see the tarball farmers. But they were tiny chess figures way down the beach. It's odd to say on a beach where hundreds of men wielded shovels and trash bags, but as I stopped to crack a beer, I was filled with that sense of solitude and euphoria that has always drawn me to the coast. I moved away from the workers and thought that maybe I had reached a place where I was relatively alone. That is until I noticed a woman walk down to the water before circling back to a hummock. When she reached the hump of sand she bowed her head as if in prayer. I walked over to say hello but then I saw the flowers and realized that she might be spreading the ashes of a loved one.

Or she might have simply been mourning for the beach itself. Either way, I gave her another mile of space before finding a spot tucked into the dunes. There I took out my journal and binoculars and studied the surf and sky. I watched a single tern dive, slashing downward and then lifting back up into the air. Success! It flew upwind with a sliver of minnow in its mouth, heading back toward its colony in the dunes.

So this is our national sacrifice zone, I thought. It was my first good view of the Gulf of Mexico, whitecapped and windblown, and of course just looking at the water was not enough. I stripped down to my boxers and walked to the edge. Then, after some brief hesitation, I dove in and began swimming out into the sea.

"If it looks clean you can go in," says BP. As if anyone, even the diving tern, with vision six times better than ours, can see the quality of the water it dives into. As if anyone, even the tern, can possibly peer into the fish it holds in its mouth and see the gift of chemicals inside, chemicals perhaps already doing their ugly work. As if it can discern that once again human beings somehow can't comprehend the simplest of notions, one the bird knows deep in its hollow bones: that everything in the world is connected, and that when you soil one thing you soil it all.

Unlike the bird, I had some idea of what I was getting into. But I swam anyway, rising and falling with the waves. It was an unctuous baptism but a baptism still. I had come a long way and now I was part of it, and I knew, as my philosopher-waiter reminded me, that we were all part of it. I also knew this: There would be no more sitting on the sidelines. If the tern was going down, so would I.

And while the water might be poisoned, for the moment it felt good. I am glad to be here, I thought. There is no other place to be.

I. In the Thick of It

THE GREEN SUN RISES: JULY 16, 2010

I've been down here ten days now. From Fort Pickens I traveled to Mobile and from Mobile to Bayou La Batre, fictional shrimping grounds of Forrest Gump and real shrimping grounds to hundreds of now-unemployed men, and from there on to Mississippi and Louisiana. I keep thinking things can't get any stranger but then they do. Everyone is either angry or giddy-drunk with the money that BP is handing out to assuage their bile. This afternoon I saw New Orleans for the first time and was tempted to stop for a drink and some *pommes frites,* but blew right through, intent on getting down here, a hundred miles south of the city. I am edging closer to the source. As I left New Orleans I couldn't help but feel that I was driving *downhill* as well as southward, even though the city itself is, in places, *below* sea level.

The nearer I get the more obsessed I become with the need to see the rig, the oil, the whole ugly mess of it. It is dark, past nine, by the time I finally close in on the town of Buras. I am feeling a little paranoid, and not just because I am foolishly sipping the beer that I hold between my legs as I drive. Hardly any vehicles are on the road but the few that are seem to be almost exclusively cop cars. I toast them in the rearview after they have passed, acting bold but feeling scared. In the darkness I can see nothing other than the road, but I sense that the land is growing narrower, and that big water is closing in from both sides. The place *bursts* with water,

water hungry to swallow the thin and tenuous land. On my left is the Mississippi, straining against the earthen mounds of levee, and on my right are millions of acres of wetlands. Straight ahead lies the Gulf and the blown *Deepwater* rig.

Following instructions, I pull a U-ey after the giant flagpole and circle back to the Cajun Fishing Adventures lodge where I will stay. Inside, the cavernous lodge is almost empty. Empty except for Lupe, short for Lupida, the Mexican cook and caretaker, and a small group of men and women who look less like fishermen or hunters than a band of scruffy Ultimate Frisbee players. The lodge's emptiness seems odd, given that I noticed several No Vacancy signs on the way down, but I forget about this when Lupe hands me a glorious sandwich stacked high with turkey, mayo, tomatoes, and lettuce. I wolf down the sandwich with a beer from my cooler, and before long Lupe and I are chatting away despite the fact that she barely speaks English and that my Spanish is comically primitive.

When we are done talking, I walk over to say hello to the other lodgers, approaching the only one who seems close to my age: a pretty woman who tells me her name is Holly. At first I can't figure her or her crew out: clearly they are not like the folks who usually populate this place, burly men who come down from Michigan or Indiana to hunt and fish and chew and spit. Not only are they a lot younger than me, but their general vibe says "California."

The mystery is solved once Holly tells me that they are part of Jean-Michel Cousteau's film team, here to make a movie about the oil spill and its effects on the sea life. As a kid, I was a big fan of the great Jacques Cousteau and loved his undersea adventures, and now I am more than pleased to be in the presence of his virtual descendants, a team of divers and cameramen. Since Jean-Michel is not here, Holly is in charge, and after we chat for a while, she introduces me to another member of the group, a young cameraman and scuba diver named Brian.

Brian and I hit it off right away. He is low-key with a ready sense of humor, and before long we are swapping war stories from our time so far in the Gulf.

"We were the first people to dive down and film what was happening underwater," he says. "We dove right into the oil. When we got back the neoprene on our diving suits had bubbled up. It looked like it was curdled."

We talk for a while more and then Holly, unprompted, does something wonderful. She invites me to come along the next day when they will be flying out over the rig in a helicopter. I thank her for her generosity. Since my plan was to have no plan and since my contacts in the area are nil, this is a ridiculous bonanza.

To top off the night, Brian and I head out to the patio in front of the lodge and drink a couple of beers. I tell him how I felt paranoid during the drive down and mention all the cop cars. He assures me I wasn't being paranoid; it's true that everyone who isn't working for BP seems to be a cop, and cop cars lurk behind every sign and shrub. There is talk, too, of planes heading out at night to spray dispersants on the water under cover of darkness.

"Napalm" is one local name for Corexit, the chemical they spray. Another is "Agent Orange," in part for the way it stains the water and shore.

"It's a very strange place," Brian mutters, shaking his head.

As if on cue a squat little truck appears near the top of the driveway. He points his beer at it and tells me to watch.

"It comes every night."

It looks innocent at first, like it's selling ice cream, minus the tinkly siren song music. Brian says it patrols the streets after dark, happily bouncing along and spraying a huge cloud of God-knows-what.

"I assume it's some sort of insecticide," he says.

It turns in at the lodge and rumbles down the driveway, as if to make sure it sprays us where we sit on the porch. We sip our beers and stare as the little toxic ice cream truck trundles by.

～～～～

I wake at five and decide to drive down to the southernmost tip of Louisiana, the very end of the land. A chronic early riser, I make a cup of coffee and throw my telescope and binoculars in the car and head down, expecting the road to be deserted. But within minutes I am caught in a bizzarro-world rush hour on a too-dark, single-lane road, the cars practically bumper to bumper. I am driving south in hopes of seeing some birds, but the other drivers on this morning pilgrimage have very different goals in mind. Some are local but many have come from far away, beckoned apparently by the smell of opportunity that often wafts up from disaster. They're headed to a harbor where they will climb aboard a motley collection of ships that includes pleasure boats, shrimp trawlers, and charter fishing boats—all called "Vessels of Opportunity," the fine Orwellian name dreamed up by BP. *VOOs*, as they are known locally, are the ships hired by the oil giant to search for oil slicks and lay boom, a kind of absorbent guardrail, to stop the oil's advance. Down here money is suddenly gushing along with oil, though not everyone is getting in on the fun. A few of the boat owners have managed to get rich by earning a couple grand or so a day to have their boats sit idle, as backups, giving birth to another new local term: "spillionaire." But most of the local men are simply struggling to make up for lost income, lost because they can no longer make a living catching fish or shrimping or trawling for oysters.

Hundreds of cars pour south toward the harbor, all going from roughly the same origin to the same destination, from their homes and hotels in the north to the harbor in the south, but none of them doing anything as unmanly as carpooling. "You should see it on a weekday," says the guy buying a tin of Skoal at the convenience store. I ask him if he's from here and he says he's not. He's from Texas and is staying in the barracks-style hotel down the road that is putting up a lot of workers.

The cars crawl south for another mile or so before turning through a gate in a chain-link fence topped with barbed wire where two guards are posted. I pull over on the other side of the road as the commuters report for duty at what looks like a military installation. Paranoia fills the air here, thick as the humidity, and as I watch the workers park their cars, I also keep half an eye out to make sure no one is watching me.

It's still the bluish dark of early morning as the workers trudge over to their Vessels of Opportunity. Street lamps spray down unnatural aureoles of light as if putting the men on stage. I know I'm in the thick of it now. Proof of that is the sign across the street that reads "Halliburton Road—Do Not Litter." Good advice. The men climb aboard their boats. One of the small, sad sights down here is watching the boat captains, seamen who have likely not worn life preservers since they were toddlers, all buckled up in their vests as they putter out to sea each morning. It looks like a badge of shame, which of course it is, beholden as these men are— not to their own government, but to the liability lawyers of a multinational corporation. The life jackets are just a physical manifestation of an ugly fact: when you sign on with BP you also sign away the right to criticize the company.

It's hard to explain to someone who hasn't been here how pervasive BP's presence is. I think of a talk I had with Ken Heck, a scientist who works at the Dauphin Island Sea Lab in Alabama. Heck has been commissioned by the National Oceanic and Atmospheric Administration (NOAA) to help figure out how to deal with the effects of the spilled oil on sea grass. This is challenging enough in and of itself, but made more so by the fact that BP has been included in all the discussions the scientists have had.

"It's kind of paradoxical," Heck said. "The way it's set up, BP is involved in every step of the scientific process. That means they know the problems we are having and what our weaknesses are. They're in on the conference calls, so let's say I mention that we really don't have any good pre-oil data from this one coastal area. Well, now they know this and later they can use this when we press for damages and they say 'Not a valid claim.'

"This is particularly troubling when it comes to damage assessment, which is mostly what we are doing now. They're not just in on every phone call, they're out in the field with us. Their representatives came with us when we went to take pre-oil samples on the Gulf Islands, before the oil got this far. They watched us like hawks. You had to put on plastic gloves every time you took a sample to avoid cross-contamination. Well, one time we forgot to change the gloves. And you know they're noting that and that later, when the divorce comes— and everyone knows the divorce is coming—they will say, 'Well, isn't it true, Dr. Heck, that you didn't change your gloves on every sample?'"

Now, as the VOOs putter out to sea, I find myself irked by the fact that a whole region is beholden to a *company*. I

hadn't anticipated feeling this way. I am no Libertarian and
I won't be attending any Tea Party rallies anytime soon. But
what about the original Tea Party? What about our auton-
omy and independence and responsibility to our own citi-
zens? Questions bubble up. How can environmental groups
and scientists be reporting to a British oil company? Are
we really buying this crap? I can't quite get my head around
the fact that BP's representatives are out on scientific survey
boats, noting facts that might be useful as evidence in some
future lawsuit. Or that their minions have been allowed to
run the show at Fort Pickens, which is, after all, a *national*
seashore.

Almost everyone along the Gulf seems to have signed
a deal with the devil, a devil that in this case isn't repre-
sented by horns and pitchfork, but by BP's green and sunny
logo. How can so many of our organizations, scientists,
fishermen, and workmen be working for a foreign corporation? How about someone sensibly saying, "Hey guys, I
don't know about you, but it seems kind of wrong to me
to hand over so much power—so much of our, excuse the
word choice here, *freedom*—to a foreign corporate entity,
particularly one that just soiled our waters and coasts." Sure
they should pay for the mess, but here's an idea: what if we
bossed them around and not them us?

I'm feeling worked up as I drive farther south, but the
place quickly pulls me out of my overheated head. Soon I
am splashing through three feet of standing water where
the river has rushed over blacktop. My headlights flash on
impromptu wetlands that cover the road. I observe a black-
crowned night heron along the edge of the water. My car
sloshes through the overflow until I reach a small, rundown
marina where a sign says, "Welcome to the southernmost
point in Louisiana." I find a spot beyond the fish-scaling table

covered with old screws and rusted bolts, between some
weeds, paint cans, and a midden of empty Bud Lights. I set
up my lawn chair and telescope at the very tip of the land, the
southernmost of the southernmost, and rest my now-cold
coffee on an upside-down white plastic bucket.

It's still half-dark, but I can make out a partially sunken
tugboat that looks like it never recovered from Katrina,
and the birds, of course, which are suddenly everywhere
as the sky lightens. A green heron hunts from the dock, a
half dozen more white ibises skirt an oily puddle, and egrets,
splashes of white, dot the trees.

This is ground zero for the spill, or at least about as close
as you can get to ground zero on the mainland. The oil is
spewing some fifty miles across the water from where I stand.
Yesterday, on the drive down from New Orleans, I pulled
over and parked next to the water and called Rocky, my
contact for the environmental magazine I am on assignment for, and explained that I wanted, *needed*, to get out
on a boat. He replied calmly, "I can probably get you out on
a boat tomorrow or the day after." He did not understand.
My world was not calm. Things were crackling in my head
and I needed to seize the moment. I needed to get out on
the water *immediately*. I tried to explain this. Though he still
didn't seem to get it, he gave up the name of a local charter
fisherman.

Captain Sal's line was busy but when I finally got through
he agreed to take me out for a short ride, warning that some
thunderstorms were coming in. Half an hour later we were
pulling out of the Myrtle Grove Marina and down a canal
in the bayou, heading toward the Gulf. As we flew across the
water I saw my first oiled pelican. It was black and flapped
heavily in front of our boat.

"There are too many rules about the oil," Sal said, shaking

his head. "We were out looking for birds and we saw a pelican sitting on some boom and we pulled up to him and the guy I was with grabbed him. If the bird had got over the boom he would have died. He was too oiled up. So we get him in the boat and then we call the hotline for the oiled animals. The girl on the phone says: 'What's your nearest cross street?' And we say we ain't near any streets—we're on the water. And she says 'Well, what's the closest restaurant nearby?' Well, I say, there are no restaurants—we're on the water. So finally we get hold of the wildlife rescue people. They come up in their boat and meet us out on the water. We reach the bird out of the tank where we'd put it and they say 'Whoa, whoa, don't touch the bird.' They put their white space suits on and their masks before taking the bird. All the while the poor bird is suffering. So we say 'Hurry up. You don't need the suits—just wash your hands after.' And they tell us that if any of us are caught handling the bird the authorities could shut down our whole operation and fine us. They would rather have the bird escape and die than get in trouble for helping the bird."

After a while, Sal and I made our way to the outer fringes of marsh. It was there that the oil first struck in large quantities, the great wave of it darkening the fringes of this immensely green and vital landscape, turning it into something dark and necrotic. What I saw was black and burnt, to the point where, if I hadn't known better, I would have thought it was the result of a small forest fire. The place looked devastated.

"Erosion is what is killing us here," said Sal, pointing at the black fringe. "And when the oil hit we got about five years of erosion in one night."

It wasn't a pleasant sight, but it was good to see with my own eyes. Even before the oil started to gush, I had started to connect the dots between our need to consume and our intensified storms; between our rising water and our use of fossil fuels; between the destruction of the wild places we love and our hunger to exploit the energy in those places. I am not alone in making these connections, of course. We are all vaguely aware that our gluttonous ways are unsustainable, but we've also got our lives to live, thank you very much. And yet. There, staring at the burnt marsh, it was harder to pretend that everything was hunky-dory, harder to pretend that we can skip through our lives with no consequences. There, staring right back at me, was the dark result of our choices.

I am not that hopeful about our ability to change. But this is not about hope. It is about looking a thing in the eye. It is about keeping an honest ledger sheet. It is about adding up what is lost and what is gained. Are we so desperately hungry for this one particular type of fuel that we are willing to sacrifice our beautiful places, our homes, in a desperate attempt to slurp up what is left? Maybe the answer is yes. But if it is, we can at least do the math with open eyes. What are we getting and what are we giving up? If this is really our national sacrifice zone, then we had better figure out just who or what is being sacrificed and who is doing the sacrificing. *Sacrifice* is a tricky word, and as a verb, it cuts both ways. It's also a broad word and many things, from paying more for twisty lightbulbs to sacrificing an Aztec virgin, fit under its tent. So far it has also been an ineffective word, with most of us turning our backs on the notion that anything really has to change.

We are all happy enough with the idea of sacrificing as long as that doesn't involve sacrificing anything ourselves. But at some level we all know something has to give. Know it but don't want to see it. Maybe the first step is seeing honestly, which means at least owning up to what is really happening. Looking it in the eye. The good thing about being here is that I can't help but face it now.

Captain Sal pointed at the feeble signs of defense against the oil. The boom looked like a child's flotation device, the pool noodles my daughter uses, only hundreds of them lined up. If they looked frivolous, their job was not: to corral and keep oil out of fragile wetland ecosystems. Lately they had been trying out a special white absorbent boom, which, Sal told me, was locally called "tampon boom." It floated ten yards from the grasses, while the newest brainstorm, boom-like cheerleader pom-poms, had been spread over one marsh island, apparently with the hope that the many cotton tentacles of the pom-poms would absorb better than the single-limbed boom.

"When the oil first came in it was the viscosity of peanut butter," Sal said.

It was still possible to see its effects—most obviously the burnt look along the marsh edges—but we saw no actual oil. Sal thought this was due to the dispersants.

"They must have upped the nightly dosage," he said, shaking his head. "We won't know the real effects of this for years."

Still, it was beautiful out on the water. Storms were coming—we could see them both on the GPS and with our own eyes—and a pink hue lit up the sky where a fingernail clipping of a moon hung. The undersides of the high clouds burned a reddish pink and when the lightning hit the whole sky turned electric. Despite the brewing storms, Sal said this was his favorite time to be out on the water, and I agreed. But he decided to head back in when he saw a waterspout—a small watery tornado that rose up out of the ocean—to the east. It looked quite beautiful but could reduce the boat to splinters in seconds.

"That's nothing to play around with," he said.

I noticed that the clouds on both sides of us had darkened.

"Are we between those two different storm clusters?" I asked.

"Actually about five," he said, gesturing down at the GPS. He pointed to a particularly large cluster. "And that one's chasing our ass."

Though we did our best to outrun the one behind us and skirt the others, the skies opened when we were about half-way back, leaving us soaked through.

As of this morning, the *Deepwater Horizon* rig has been gushing for seventy-seven days, filling the waters in front of me. Some think of the Gulf as our least coast, the place where we dump all our shit. That shit includes hundreds of other leaky wells, as well as nutrients, waste, and fertilizers that are carried down rivers from Midwestern farms and emptied in its waters. No matter how we try to dress it up, it's an afterthought, a dumping ground, the country's toilet where we flush our waste. Now added to the mix are, at most recent count, about forty million barrels of oil and millions of gallons of dispersant. The president just called this the "worst environmental disaster America has ever faced."

And yet, if the Gulf is a hell of sorts, it's a beautiful hell. I walk up the road and stare out at three cypress trees that must hold a couple hundred roosting ibises, all of them set-tling and fidgeting, settling and fidgeting, like fussy sleepers. Beyond the ibises, I catch sight of a roseate spoonbill, an anomalous patch of bruised pink in a green cypress.

I am here to see what is left, and part of what is left is beauty. I need to be honest about that too. I have grown weary of avoiding things; it takes a lot of energy. Bring on the beauty and the ugliness both. Show me the honest math. What are we losing? What is being gained?

The birds have already lit up the trees, and now the orange ball lights up all of it, the world greening as it brightens. There is a vibrancy here that reminds me of a word I learned from a former *chiclero,* a Belizean man who helped find and tap rubber trees. "Yax" is a Maya word meant to describe a particularly vibrant and wild green. Here there is *yax* aplenty, from the cypresses to the marsh grasses and ferns to what I take to be some sort of elder plant. In counterpoint stands the rising sun. Despite what must be considerable pressure to sell out, it still rises freely, brandishing its usual blazing reds and oranges, not yet willing to don the corporate colors of green and yellow. It is nice to know that there are still a few realms beyond the reach of British Petroleum.

I sip my cold coffee in salute to the sunrise.

FORCES OF NATURE

When I return to the lodge from my early morning bird-watching I get some disappointing news: the weather has put the whammy on our helicopter ride. We will not be getting up in the air today. I sulk around the lodge for a while like a child denied a ride on a roller coaster and think of what could have been. Then Lupe comes to the rescue with a pitcher of ice tea.

Somewhat buoyed by the drink, I step out for a walk in the drizzle. It's no helicopter ride, but it brings unexpected sights. I head instinctively toward the river, first hiking up the levee and then walking along the top. From here, on the hump that sometimes struggles to contain it, the river looks muddy, caged-in, powerful. I've heard lots of songs about levees, of course, but didn't grasp the concept until now. Walking on top gives me a view not just of the river and the opposite bank, but also, to the west, of the wetlands that lead to the Gulf beyond. It occurs to me that I have never walked along the banks of the Mississippi before. Which seems an amazing fact, since I happen to be an American.

After a while the rain stops and the sun breaks through. The heat is overpowering. It slams you, stuns you, slathers you in sweat. Everything wilts, and I am part of everything. It's the kind of heat that makes you want to lie down and give up, to throw up your arms in surrender. It helps you understand the logic behind siestas; every instinct telling you to crawl into a cool, dark place and lie there and be still. The heat even seems to stun the birds that fly overhead; they flap lazily and deeply.

Crickets blare and willows sag and down by the water the reeds grow as tall as trees. By the time I get back to the lodge, I am sopping with sweat. I find Holly at one of the high tables along the rim of the lodge, sitting on a barstool and typing on her computer. When I walk up, she tells me she will soon be interviewing a member of an organization responsible for surveying birds and counting their fatalities from the oil. She invites me to sit in during the interview. It turns out to be an odd and frustrating exchange that takes about forty minutes to go nowhere. Holly is gently prying and persistent, but no matter what she says the man will not divulge the number of bird deaths. Finally, sheepishly, he admits why.

"BP is now on our board of trustees," he says.

The interview ends soon after. Once the man has left, all we can do is shake our heads and laugh.

The rest of the day passes quietly. The Cousteau folk are working on their computers and lying low. I decide to take a nap. But the quiet ends with the arrival of Ryan Lambert in early evening. He walks into the lodge as if he owns the place—which, of course, he does.

Not only does he own it but he nailed together almost every board with his own hands, or more accurately, renailed them after his lodge was drowned by Katrina. We shake hands and he points up at the rafters above the mounted animal heads, to a line in the wood over twenty feet up.

"That's how high the waters from Katrina reached," he says.

Soon we are all sitting around one of the long camp-style dining room tables while Ryan holds court. He takes the head of the table, while Holly and I sit on either side of him. The

Cousteauians and I don't say a word, which is fine with me. Ryan is a big man, not especially tall but burly, and you just know he has told a thousand fish stories to paying guests from his seat at the head of the table. He has huge hands and a big expressive face, red from decades in the sun, and he looks like he could pick you up and crack you in half over his knee. But more impressive than this implication of physical strength is the immediate impression of energy, his pilot light always on high.

"I'm the only lodge around that isn't booked up," he says. "The rest of them are filled with BP workers. But I'd rather meet interesting people than whore myself out to BP."

Over the next hour I learn that Ryan was born and raised in Louisiana, just outside of New Orleans. His grandparents owned a place down here in Buras, where they rode out Hurricane Betsy. When Ryan visited as a kid he would roam the wetlands, hunting and fishing with his uncles and falling madly in love with the place. After high school he went right to work at a chemical plant, but he couldn't get this wild place out of his head. After a few years at the plant, he decided to start doing the impossible. That is, he kept working at the plant full time each night and then drove down here to work as a fishing and hunting guide about twenty days a month. Which left about ten nights a month when he actually got some sleep. He kept this going for twenty-one years. Finally, at thirty-nine, he quit the plant and moved down to establish Cajun Fishing Adventures, which grew into a million-dollar business with over twenty employees including fishing guides, duck guides, house cleaners, and cooks. He had realized his dream, at least until Katrina struck.

"I've had a bull's-eye on my back for a while now," he said. "First Katrina and now this."

He insists that I join him and the Cousteau crew for dinner, and when I tell him that dinner is not included in my deal, he laughs and waves it off. Lupe serves up spareribs and coleslaw while Ryan tells the story of how he rebuilt the lodge after Katrina. The others lean in to hear.

"I got a plaque for being the first person to come back to this parish. I came by boat at first. It was a watery wasteland where you could only see the peak of this lodge. Everything was dead."

He wanted to rebuild as soon as the water receded, but the insurance company refused to pay him, claiming the damage had come from water, not wind. He needed money so he came up with an idea. There was talk everywhere of trying to revive local businesses and of cleaning up after Katrina. He added these things up and put together a crew to clean debris and rebuild the parish. He worked hard and was paid well, well enough to in turn pay for the materials to rebuild his lodge, which he did whenever he wasn't working on the cleanup. "I was possessed," he says. Though I have only known him about an hour, this is not hard to believe. About a half year later, Cajun Fishing Adventures was up and running again.

"And now this," he says, shaking his head. "This is worse than Katrina."

Considering that Katrina basically flattened his town before drowning it when the Mississippi broke through the levee, I can't help but ask: "Worse?"

"Worse, I think. Not just the oil but the dispersants. I was out on an island the other day and hundreds of small clams were rolling in with the surf, all of them covered in tarballs. The dispersants have sunk the oil out of sight of the cameras, but it's down near the bottom of the ocean, at the base of the food chain. This is just the start of the death we're going to be seeing in the future. The fisheries were already dying. This could be the deathblow."

I have heard this before tonight, but never put so bluntly. While no one knows how the chemical dispersants will affect the Gulf's food chain, everyone is anxious. The original mixture BP proposed using on the spill—Corexit 9527— was deemed too toxic by the EPA. When they demanded that the company change to a less toxic product, BP simply switched to another version of the same chemical compound, Corexit 9500. Both versions of the chemical are manufactured by BP and neither are legal in England. At this point, over a million gallons of the stuff has been dumped in the Gulf.[2] The fact that the EPA did not even attempt to enforce its own ruling says worlds about what is happening here. The immediate effect of the chemical is to first disperse the oil, and then to sink it to the ocean floor. This makes short-term sense to anyone, like BP, who wants to tamp down immediate panic about the spill, since it means that less oil will be washing up on beaches in a region that depends on tourism more than any other industry. The goal is to not have this look like the *Exxon Valdez,* which is to say the immediate goal is focused on appearances.

"It's like a kind of magic trick that BP is trying to pull off," Ryan says. "A sleight of hand, out of sight out of mind. It's good PR if people don't see the oil." Whether it's good for anything else is a question he leaves unanswered.

After a dessert of blueberry pie, we retire to the overstuffed couch and turn on the Weather Channel. Ryan sprawls out on the big pillows and then asks me what I found on the beaches of Florida and Alabama. I tell him and then he asks for an overall impression.

"Everyone's got their hand in the till," I say after a minute. "That's what I've come to believe. Everyone's a part of it."

"I'm not," he snaps.

I'm afraid I've offended him, but it turns out Ryan Lambert

is not easy to offend. He is smiling now. His reply was not defensive. He was simply stating a fact.

After a while the others head to their rooms or to the various high-top tables around the big room's rim, and Ryan and I stay up late talking. To an outsider we might seem an unlikely pair; and it quickly becomes clear that in many ways we are at the opposite ends of the political spectrum. As simplistic as it may be, he's the perfect stereotype for the hunting enthusiast, salt-of-the-earth conservative, and I, the nature-loving, college professor liberal. And yet we both possess an underrated quality often ignored in the televised and shrill national debates: a sense of humor. More importantly, it turns out we have birds in common.

Over the last fifteen years or so, my life has gotten tangled up with birds, and specifically a particular species of bird called ospreys. I spent an entire year on the marshes of Cape Cod observing them, and another migrating with the birds down the East Coast to Cuba and Venezuela. Majestic birds with six-foot wingspans and black raccoon masks, ospreys get their living by making high dives into the water for fish. When I heard the Gulf was filling with oil my first thought was: the ospreys will be diving right down into it.

Ryan is a bird lover, too, though of a slightly different sort. While I watch birds through binoculars, he likes to shoot them.

"The only difference between conservationists and environmentalists," he says, "is that we eat our way through nature."

Ducks, not ospreys, are his obsession, but we are both worried about the fall migration. As Ryan sinks deeper into the couch, he describes the spectacle of millions of waterfowl sweeping down through the marshes of Louisiana. He is a tough guy, and you have to remember that what he wants to do with these birds is kill them, but his voice softens when he talks about the coming migration.

"This estuary is the richest in the nation and the majority of the waterfowl in the United States will come through here. They come through in great waves. In late August the blue-winged teal will come through. They feed on the bottom where the oil is. They are a beautiful bird, meticulous, too, with never a feather out of place. When they preen they will spread the oily mousse all over them and when they get oiled up they can't regulate their body temperatures and then they can't fly. They will not be easy to find and clean. You can't catch a duck the way you can a pelican. They're elusive and can get deep in the grasses or underwater. You won't be able to catch them. Not the scaup and redheads and canvasbacks. The numbers of these birds are down already. And now they will be living out in the oil."

At one point I excuse myself and run into my room. I dig into my bag and find the book I'm looking for, a compendium on migration by Scott Weidensaul called *Living on the Wind*. Returning to the couch, I read out loud a sentence that I underlined just yesterday: "Migration depends upon links—food, safe havens, quiet roost sites, clean water, and a host of other resources, strung out in due measure and regular occurrence along routes that may cross thousands of miles. But we are breaking those links with abandon."[3]

He doesn't say anything at first and I'm afraid I seem ridiculous—a liberal, book-quoting caricature—compared to my manly host. But he doesn't laugh at me. He nods and seems to chew over what I have read.

"There is no bigger haven than this delta," he says. "Think what we're destroying. Millions of acres of wetlands. If we lose this we lose everything."

Millions of other birds, not just blue-winged teal and

ospreys but a hundred other species, will pour down this central corridor as they make their arduous journeys from points north to Central and South America. Migration is always a gambit: everything has to go right. It is a time of both stress and opportunity, but in this strange and oily year, I worry that the former will overwhelm the latter.

It's late by the time Ryan heads back to his house, which is just a few hundred yards from the lodge. Holly and a couple of the others are still working on their computers, but I keep to myself, nursing a beer and chewing over my talk with Ryan.

I think of the thousands of ospreys that will be heading this way soon, when the weather changes. Ospreys have become no less than a way of organizing how I think about the world. A lot of nature lovers pay lip service to trying to imagine the world beyond the human—the biocentric as opposed to the anthropocentric—but by getting to know one animal well, I have, almost despite myself, done just that. My thinking, more specifically, has become *osprey-centric*. The birds are the one thing in the universe I pick up and find everything hitched to.

And so, when I think about the millions-plus gallons of chemical dispersants being dumped in the Gulf, my mind keeps returning to ospreys. We don't yet know what the long-term results of Corexit 9500 will be on fish and birds, and we may not truly know for years, but we know that both the oil and the chemicals are already deep in the water column and may soon pervade the food chain. If this happens it will not be the first time ospreys have been impacted by human tinkering. In fact, if the birds were aware of what was happening in these waters, they would no doubt be thinking the osprey equivalent of "Jeez, not again." This is a species, after all, that was all but eradicated by chemicals in the recent past.

It was from learning about DDT and ospreys that I first came to understand the concept of interconnectivity. The story begins in the late 1950s when DDT was sprayed on fields and marshes with the goal of eliminating disease-carrying insects. But what the chemical proved, in a giant science experiment not so different from the one currently going on in the Gulf, was that "the web of life" is not some fanciful notion that a groovy ecologist invented. In fact, the way that DDT moved through that web—killing the insects but also moving up the food chain to vegetation and smaller fish, accumulating in larger quantities with every step up, eventually settling in lethal quantities in top predators like ospreys and eagles—was almost as miraculous as the web itself. Almost. But while the web created life, the chemical brought death. The way it killed was particularly cruel: it caused a thinning of the eggshells so that when parents sat atop their eggs to incubate them, their offspring were crushed. Before the chemical was banned, the birds were almost entirely wiped out in the Northeast.

And now we are at it again.

The people who made and sprayed DDT were not evil. Who wouldn't want to get rid of mosquitoes? They weren't evil, but they just believed that they could control things. They believed they could make things better than they are; that they could always fix what got broken; never considering that some of the things they were breaking had taken a million years or so to make.

As tragic and awful as the oil spill is, the use of dispersants could prove worse in the long term. You can at least argue that the first mistake, the spill itself, was an accident, an accident born of arrogance and greed and oversight, but still an accident. The second mistake grew out of opposites:

conscious decision and panic. A friend who works with the BP administrators in Mobile told me that during those first weeks everyone's eyes were wide from fear. Fear has quickly led to a desperate need for the illusion of control.

I don't claim to know what Corexit or other dispersants will do, how exactly they will infiltrate and affect fish and birds. But I know that good science is born of skepticism, and that those who confidently claim that they *do* know are exhibiting the thinking of little boys. By this I mean a rambunctious, occasionally effective, headstrong, and insistent way of being. I mean a demented can-do philosophy that wants wants wants and so will find a way to get. I mean a way of being that most of us, boys and girls, eventually grow out of, a way that can actually lead to the building of some pretty neat-o things, but that also leads to pouring boiling water on ants. If you think I am exaggerating, if you think that this isn't the kind of thinking running rampant down here, then consider this: last week a number of people started discussing the possibility of detonating a nuclear bomb to plug up the well.[4]

But forget nuclear weapons. Who needs them when you have Corexit? How do little boys deal with things they break? Sometimes they hide them. You break a lamp and don't want Ma to see it so you put it in the closet. The ocean floor is now the closet. If the Gulf is our national sacrifice zone, then the ocean floor, where life starts and where the dispersants are sinking, will be the Gulf's own double-secret sacrifice zone. Talk about sweeping something under a rug.

To a lot of fairly knowledgeable people, the spraying of dispersants seem as much a disguise as a solution. "A magic trick," Ryan called it. And a good one. "Out of sight, out of mind" could be our national motto. We put so much energy into pretending, into avoiding, into not seeing what is. And what are dispersants if not a way to hide reality? A way to make a problem appear to be gone. In this regard they are an embodiment of both our belief in the importance of appearances and our own unwillingness to acknowledge oil—both oil in general and oil in the Gulf specifically—and its consequences. In other words the perfect solution, from a poetic if not practical point of view, for a society that doesn't want to face its own reality.

The problem is that it's hard to sweep things under the rug in nature. Things insist on being part of other things. Ospreys know this. What DDT taught is that the invisible can kill, and all our denials and disguises and PR moves don't make a thing any less lethal.

~~~~

Luckily my trip so far has been filled not just with osprey ideas, but with actual ospreys. On that first night in Fort Pickens, I camped next to a forest of dead trees. The trees, killed by the salt from Hurricane Ivan, twisted up into the sky like mushroom stems. The hurricane had created perfect homes for ospreys, leafless branches allowing panoramic views of the dunes and water. In their crowns sat many great shaggy nests.

On the beach beyond the dead forest I watched a single osprey as it hunted. It hovered above the crashing waves, its black and white wings semaphore flashing. I first came to bird-watching as a sports fan, craving contact, and that day didn't disappoint. Soon enough the osprey dove, pulling its wings into a *W* shape and hurtling toward the water. Ospreys don't twist and plop into the water like pelicans, or dart down in the manner of terns. They *dive*.

As it happened, the bird missed on the first attempt. 0 for 1. It shook off in the air like a wet dog, shivering, and then tried again, but pulled up at the last minute, as if it were all a feint.

"No fish."

I said this out loud without thinking about the larger repercussions of my simple caveman sentence. But. What if there really were no fish? I knew that this particular bird needed three or four a day, more when feeding a family.

Later that evening at sunset, I drove into the Fort Pickens picnic area and found a spectacular nest. The picnic area had become mission control for the cleanup efforts, and the birds now shared their habitat with trucks and Dumpsters and fluorescent-vested workers and dozens of Porta-Potties and hundreds of all-terrain vehicles. Still it was the perfect place to end the day, and not just because the nest was one of the most beautiful I'd ever seen, a nestled cup of sticks in the upper branches of a dead live oak. Three young ospreys— immatures, identifiable by both their size and checkered wings—perched around and on the nest, illuminated by the last shafting rays of sun. They let out high-pitched warning cries that told me not to come any closer.

Then I heard another sound that took a minute to recognize. The osprey cries were mixing with a different sort of music: the backward beeping of trucks. On the far side of the parking lot, air-conditioning blew through ducts into a huge billowing tent and men in Hazmat suits walked in and out. It looked like a scene out of Spielberg: the military trying to keep the discovery of the alien autopsy under wraps. But what might have been too clichéd for a movie was the fact that above it all the ospreys nested, the whole scene watched over by birds that had come back from the dead.

II. My Crowded Day

UP IN THE AIR

Today the weather cooperates. The helicopter arrives, landing on Ryan's lawn at eight in the morning, splattering shadows outward. I run toward it, hunched down and sensibly worried, like any beginning helicopter passenger, about decapitation. While I've never been in a helicopter before, Brian and his coworker on the Cousteau team, Nathan, are old pros, having been out to the rig not long after it first blew. I, on the other hand, am like a little kid, clutching my disposable camera and notebook as we lift off. The blades spray wind and light across the grass below. A wavering turkey vulture flaps away from the noise. The lodge shrinks as we rise, and I see that my first impression—this land as a fragile strip between waters—was correct. Brian points down at a cop car hiding behind a tree. As we fly south toward Venice where we stop to refuel, it's hard not to notice how much oil is being used to help clean up the spilled oil: helicopters are coming and going constantly from the pads. Once we are back in the air Brian points out the window toward a larger Black Hawk helicopter carrying a sandbag west over to one of the threatened islands. The huge bag hangs and sways below it on cables like a spider's egg sac.

Thirty percent of domestic U.S. oil production now comes from the Gulf, and in the headlong rush to drill deeper and find more, that production has increased 34 percent just since 2009.[5] If this began as a little boy mess, it has bloomed into an emergency that, according to the little boys, only they can fix. And so now helicopters fly all day long trying to save us from the oil, burning thousands of gallons of gas in the process. Everywhere you look you have ships, cars, trucks, planes, and copters charging every which way to protect us.

But for the moment I give irony a rest. What overwhelms the ironic, swamps it actually, is the landscape. It was one thing for Ryan to say that 14 percent of the country's coastal wetlands span out from the road near his home, it is another to actually see them. We fly south but can't outrun all the green, the great mangrove islands and marsh grasses. As an outsider, prone perhaps to regional prejudice, I somehow accepted that it didn't really matter much that these waters were home to over four thousand oil and natural gas wells. We needed somewhere to dump our industrial complex, but how can this place be *that* place: it shimmers with green.

Suddenly, a hundred feet below me, I see black and white wings and realize that we are soaring above a half-dozen Frigatebirds, officially known as magnificent Frigatebirds. While they are spectacular, spending whole days in the air, the adjective in the bird's name has always seemed overdone to me, since they get their living in a somewhat seedy manner. They are kleptoparasitic, never deigning to stoop as low as to fish for themselves, but instead swooping down from on high to steal fish from other, more industrious birds. They seem to float far above the mess we have made, but of course they can't stay above it for long. Eventually they must dip in and steal fish that are just as dirty as they were at the moment they left the water.

In the landscape below you can see geology at work, how the Mississippi dumped its nutrients for millions of years and how the land then spread southward from the delta, extending itself in miles and miles of watery grasslands, which in turn became home to young fish and oysters and shrimp and millions of birds. Far from a shit hole, it is a wonder. Green jigsaw pieces of grass fit with blue pieces of water while a river runs through this already-watery world. A great snaking river, hemmed in still, even at this point, funneled by engineers toward the Gulf itself. It is a shocking sight: the great freshwater torrent running home toward salt. It is still almost beyond sense. The bayou world of marsh grasses and creeks and straight man-made canals is one thing, but then through them, or looking like it is superimposed on top of them, is the great brown weaving Mississippi. I have never before seen freshwater of that magnitude moving through a coastal water ecosystem.

Before I came here I boned up on the Mississippi, and read a book called *Rising Tide* by John M. Barry. Barry's book told the story of the great flood of 1927, but what really struck me was the frontispiece. If ever a picture was worth a thousand words here it was: an illustration of the branching tree of the Mississippi and all its tributaries. The picture was called "Mississippi River System." Somehow I had never thought of it that way before. When I pictured our largest river I saw it running straight down, express-style, north to south from Minnesota to New Orleans. But this picture told a different story, a story that stretched from the headwaters of the Missouri River in Montana to those of the Allegheny in upstate New York. It made it look like every creek in the United States fed the Mississippi, which isn't far off. The feeder streams and rivers were capillaries and veins and arteries, a great cardiovascular system of almost the entire country from east to west and north to south. And all of it ends up here.[6]

We fly on. I would have thought we were past any towns at this point, but suddenly a human outpost appears below us amidst the watery grasses. Through my headphones the pilot says that this hamlet, which sits at the mouth of the

Mississippi, is called Pilottown. Here local river captains take over ships returning from sea to steer them up the Mississippi—the sea captains don't know the river's currents and tides like the local river captains do. The only way to get to this little outpost is by sea or air. According to our pilot it's a wild place, renowned for its drinking and prostitution as much as its frontier remoteness.

After Pilottown, we reach the end of land. Orange and red and blue lines bubble out around the outer islands, as if a giant child had clumsily tried to trace the land's outlines. These are the lines of boom, a colorful, if ineffective, protection against the oil. As of yesterday, 3,474 kilometers of boom have been laid down. And yet, with any sort of good wind or storm, the water and oil will splash right over the boom, rendering it useless.

As important as doing something is right now, looking like you are doing something is perhaps more important. This is a lesson BP has learned well. "False hustle," was what Red Auerbach, the old Celtics coach, called it, and false hustle has become a BP specialty. Just this week the company had to sheepishly admit that they had doctored a photo from their spill command center in Houston that showed workers monitoring great banks of video screens glowing with underwater images. There it was in the papers: three passionate and concerned workers keeping their eyes on our waters. The only problem was that half of what they were seeing on the screens had been photoshopped into the image. Before this bit of trickery, most of the screens were blank.

I try to make out changes in the water's color as we fly past the boom out into the Gulf, wondering what is oil and what is not. I see great black sheens and stretches of lighter water, but I have spent enough time on the coast to know how ocean colors can change, with or without oil. I don't want to sound like an idiot, but I decide to ask Brian if I am seeing what I think I am seeing. He tells me that he can't really see much oil at the moment, and the pilot agrees.

"You wouldn't have believed it when we first came out," he says. "You wouldn't have been able to miss it then. It covered everything."

I assume that by now much of it has been sunk to the ocean floor. When I look down my eyes can't penetrate the surface, but just yesterday I read an interview with Samantha Joye, a scientist from the University of Georgia, who spoke of witnessing black plumes, many miles long, that travel deep under the water, large dead areas with no oxygen and no fish.[7]

After another fifteen minutes we reach the *Deepwater Horizon* rig itself. As we approach, the dozens of boats below look like Tonka toys gathered around the rig, as if trying to protect and comfort it. But as we get closer, it is clear the rig needs no comfort. It is emblazoned with BP's green and sunny logo and appears almost cheery, as it is no doubt supposed to look. The scene looks not just sunny but industrious, with no hint of despair.

From up here the rig may look like a toy, but it is in fact a great metal island, capable of housing over a hundred men. In broad daylight it is hard to picture the fiery hell of April 20, the night when the methane bubble blew up through the well and exploded at the platform, killing eleven men, injuring seventeen more, and sending dozens leaping off the platform into the flaming water. What was it like to take that ten-story plunge? The chief engineer said later that he thought of his wife and his little girl before closing his eyes and making the leap. Those, I am sure, would have been exactly my thoughts.

In the story being told right now the *Deepwater* explosion

was a great tragedy, but also something anomalous, an "accident," of course, a terrible accident. But is something an accident if crucial tests are skipped, if costs are cut, if warning systems are *turned off* so alarms won't ring, and if even the CEOs of Shell and Exxon—a Big Oil gang that is known to stick together—have sworn in front of Congress that the *Deepwater Horizon* well did not come close to meeting industry standards? Is something an accident if a multi-billion-dollar company, the world's fourth largest, decides it needs even greater profits, and sends a top-down directive to cut costs company wide by 25 perecent? "I'm not a cement engineer," BP's CEO Tony Hayward told Congress in way of defense, but presumably he had a few cement engineers working for him. He also said, "I'd like my life back," a sentiment no doubt shared by the eleven dead crew members and their families.

Far from anomalous, disasters had, by the time of the spill, become commonplace in the world of British Petroleum. Over the past decade the company went from the little brother of oil to one of the big guns, acquiring Amoco and Arco in the process. But during that heady rush the company's M.O. was to take risks and cut costs, safety be damned. This is not overstatement. BP has led the Big Oil league in deaths and disaster. In 2005, fifteen people were killed and 170 injured when BP's Texas City refinery blew up due to shoddy safety standards. In July of that same year BP's flagship for deepwater drilling, the giant off-shore rig *Thunder Horse—Thunder Horse!*—was toppled, seemingly by Hurricane Dennis but in fact by faulty valves hastily installed. The next year BP hit the disaster trifecta when 20,000 gallons spilled from a rusty pipeline in Prudehoe Bay on the north slope of Alaska.[8]

Which brings us back to the question: if things happen

regularly and for the same reasons, do they still qualify as accidents? Which leads in turn to the next and larger question: if we, as a country, keep acting in ways that lead to shocking events, isn't it time to stop being shocked?

Not that it isn't shocking. A twenty thousand gallon spill like the one in Alaska is a disaster. But over two hundred million gallons have spilled from the well below me since early April.

We circle the rig again. I stare down to try to see the deeper story. It was down there that eleven people were sacrificed in the name of profit. Is that an exaggeration? Tony Hayward and Carl-Henric Svanberg might be scapegoats—and fine scapegoats they are, complete with their James Bond villain accents—but what about the board of directors? And what about the system that created the board? The group and the philosophy that demanded that this company, despite earning *billions* of dollars, had to earn even more to sate them; that to do so, to provide more billions, a 25 percent cut in operations had to be enacted, even as those operations were expanding downward into new territory, 13,000 feet below the ocean floor? How were those cuts enacted? Simply and systematically: by cutting corners and skipping regulations and eliminating safety measures. Piles of money, enough to support a small town for decades, were being divided between a board made up of a dozen or so people. And yet no one could be bothered to pay a few hundred thousand on tests, nor could they abide alarms that might slow them down.[9]

Take this down to a personal level and it seems almost inconceivable. This is not the first time I've traveled this country and I am always surprised by how decent people are. But where are all those exceptional individuals in a moment like this one? Is it only in large groups that people are allowed to bury their morals? No healthy individual would ever do to their family or friends what this corporation has done to the people of the Gulf. Individuals would face immediate ostracism. Maybe it's as simple a problem as the size of the organization, or even the words *organization* and *system*. When profit is made the greatest priority and one's job—one's self-interest—hinges on that profit, simple commonsensical goodness flies out the window.

I am wrestling with these ideas and can't stem the tide of confusion. It's too much to handle all at once. In our oversimplified political discourse we talk a lot about the importance of business and growth, but we also talk a lot about freedom and individual rights. But a corporation like BP is about as individualistic as a batch of flesh-eating bacteria— there is no debate over what the collective will is: grow and profit, no matter the cost. What does freedom mean when we blindly trust that an entity like BP will not destroy the world we rely on for our health, happiness, and well-being?

We don't stop there, though. Before I came down here I watched the congressional hearings where Tony Hayward testified. A woman jumped up from the back row and waved her hands, which she had painted black, and yelled: "He should be charged with a crime!" She was quickly dragged away. Maybe most people will roll their eyes and call her a wacko, but she is right. Rather than being charged with a crime, this man's famously inept and dangerous company is being charged with running the cleanup. It is hard to imagine a culture in which this could possibly happen: not only do we trust them, but, when they err, we trust them yet more.

As we spin over this giant pool of water a graphic and slightly inelegant metaphor comes to mind. It's as if BP were a houseguest who takes a shit in your bathtub and then, loudly and boorishly, orders your children to clean it up. Worse still, he slips each of your kids a fiver and has them sign a piece of paper promising that they won't tell anyone what really happened. The truly wild thing down here is that everyone has gone along with this plan, carrying it out as if it makes sense, nodding and going about their unsavory business.

~~~~

After we fly over the rig we head toward Grand Isle, a national park at the end of one of the green fingers of land that reaches out into the Mississippi Delta. It has long been regarded as one of the most beautiful spots in Louisiana, but even from way up here I can see the brown, burnt fringe of dead grasses from the oil. This is where the caramel goo first rolled in.

But it is also still a miraculous landscape, full of fish and birds and gators (like the one we saw after lifting off from refueling). Ryan said that 14 percent of the continental coastal wetlands fan off of Route 23, but the area we are flying over now makes up closer to 40 percent. It is green and vast, wavering like a mirage below us, and for centuries it has received the gift of nutrients offered up by the great surging river. I have to laugh to think that just a week ago I would have told you that this was a second-rate coast. There is nothing second-rate about it, other than the way we have treated it.

The United States consumes 40 percent of the world's oil. About 70 percent of that is for transportation, mostly for our cars. I am not here to wag fingers: I, too, drive a car and live in a car society. We are a hungry people. I, too, am hungry. We are hungry not just for oil but for the ease it brings, and, as creatures of habit, we have become habituated to this easy, oily way of life.[10]

Oil has often been called an addiction. Just as surely as a

junkie's life leads to degradation and crime, I can see spilled on the beaches below me the results of our addiction. Here is our degradation and here are our crimes, spread over these beaches and in these waters. We debate scientific theories in our culture. You may choose not to believe that the world will warm, and while your beliefs have little to do with what the world does, you have a right to them. But what I am seeing below me is not theory. Here, in this place, there is no disbelieving or believing. Here it is right in front of you and in your face.

What good does it do to self-flagellate? Oil in and of itself, far from being "bad," is almost miraculous in its composition and effectiveness. Oil is the "solar energy" that environmentalists like me have long cried for. Solar energy from eons ago, energy sucked in and stored by plants, now long dead, that has been squeezed by earth and time, energy that we ignite to power our cars and cities and lives. And not only is oil "natural" and miraculous, it was also, for a while, a terrifically good idea. It powered things cheaply and well. One of the reasons we are having a hard time turning to alternatives is corporate resistance, but another is that it's hard to find something nearly as effective. Who would have guessed that old fossilized trees and plants could do so much so well?

To think about oil clearly we need to clear our minds of guilt and blame. Who would have known, even fifty years ago, where all this would lead? Who would have known that it would lead to wars where our young people would die? Who would have believed that we would be capable of warming our own planet and of melting our ice caps? And who would have imagined to what extent the addiction would grow, to the point where all the oil that has spilled below, the millions and millions of gallons, would only be enough to power our country for six hours?[11] If these ideas seem too numbing, then consider a more everyday disaster. Who would have known that this substance, celebrated when it came spouting out of wells over a century ago, would have led to what is possibly the most nightmarish, if quotidian, of human miseries? I am talking, of course, about commuting.

For all this, the time for oil is passing. Not for any moral or philosophical reasons, but for practical ones. There simply isn't enough to keep going. And the little that is left is hidden in places like the Macondo site down below me, places where drilling involves great risk. Do we want to rip the world apart to get those last drops? That is part of what we are talking about when we talk about sacrifice. We will gain oil. What will we lose?

The oil companies know how they would answer that question. And yet this industry, as monolithic and scary as it may seem at the moment, will topple to its knees soon enough. You can hear the industry's death rattle if you listen closely. The corporations will not go easily, just like the railroad barons. But there is no denying the fact that no matter how deep they dig or how much of the earth they soil, there is only so much left. So what do we do next? And when will we start doing it?

⁂

As we head back toward Venice and Buras the pilot points at a giant oil skimmer that he claims is capable of sinking itself. It looks gray and unwieldy, like a battleship, and it is hard to picture it dipping down into the water.

"Kevin Costner just met with the BP bigwigs over in Port Fourchon," the pilot says. "He's got a ship that separates oil and water that his company developed. It's going on the water today."

Which of course makes sense, since the sinking land and rising water below are essentially *Waterworld*. I find myself smiling despite the grimness. It's a common condition down here. If you are in the business of collecting small ironic tidbits then this is the right place for you. Take the fact that the day the rig toppled into the sea, two days after the explosions, was April 22, or as it is also known, Earth Day.[12] Or try this one on for size: President Obama announced that he was opening up more United States waters to offshore drilling, a decision that would benefit BP first and foremost, twenty days before *Deepwater* blew.[13]

The president let the press know this on March 31 but he missed a golden opportunity.[14] Had he only waited several more hours, he could have made his announcement on April Fools' Day.

THE CASE AGAINST STRAIGHT LINES

When we get back to the lodge I want nothing more than a nap. This will not be happening. The lodge hums, a nest of activity. The Ocean Doctor, aka David Guggenheim, has arrived. David is a scientist and radio personality who has traveled the world reporting on the state of the coasts, and today he is joined by an *NBC Nightly News* cameraman who circles him like a pilot fish. The cameraman, fresh off of a stint in Baghdad, another war-torn region, is South African and his words sound thick and garbled. ("He has a funny accent," Ryan says in full-on Creole singsong.) Guggenheim's brother, Alan, has also come along and we say hello as Holly and Ryan try to organize the afternoon's expedition.

In the midst of the chaos I get a phone call. At the end of my trip out on the water with Captain Sal, we came upon a row of fish camps. Through the rain I stared out at the dilapidated shacks that lined the canal about a mile from the marina. They appeared fragile, permeable even, built as they were on a watery foundation. They were unabashedly ramshackle, pieces of plywood nailed here, a screen door thrown up there, rickety docks jutting out like the tray on a toddler's high chair. I decided I needed to find a way to spend a night out in one of them. They reminded me of other modest coastal dwellings I had known, and they were right out on the frontlines of the oil. I ended up talking with a woman named Leona, who ran the Myrtle Grove Marina store. She told me to call Anthony, a sixteen-year-old local kid who liked to hang out at the marina and whose parents owned one of the fish camps. I got his outgoing message, a loud and blaring country song, and wasn't sure I'd gotten through.

"Hallo," he says now. "Are you the guy who wants to go out to camp?"

I am, I tell him, and he says that he could maybe get me out there tonight. I explain that I will already be out on the water for most of the afternoon and am not exactly sure when I'll get back.

"That's okay," he says in a rushed musical mumble. "I'll get the place ready. I don't mind waiting."

I hang up and within minutes we are driving two cars and towing two boats across the road to the local boat launch, not three hundred yards from the lodge. The plan is to head out on the Gulf in the Cousteau pontoon boat and Ryan's single-console fishing boat. Brian and Nathan take the pontoon boat, a Zodiac VI, while I climb into Ryan's boat with the Ocean Doctor, his brother, the cameraman, and, of course, Ryan at the helm. Soon we are racing across Barataria Bay, wind and spray in our faces, and it strikes me how strange it is to be traveling through the very same waters I was just staring down at from a half mile above. Meanwhile the Ocean Doctor is interviewing Ryan, and the cameraman is bouncing on the bow, trying to get good shots of both men.

"Look at my GPS," Ryan says. "It still shows this as land. Not long ago this was 6.3 miles of solid grass. Now I can point my boat right over those 6.3 miles and never see a blade."

Ryan pulls up to a spot where wooden posts thrust up through the water. Dozens of huge black and white Frigate-birds, the same type of birds I saw from above this morning, lift off. I have never seen this many Frigatebirds outside of Central America. They rise from the posts in slow motion, beautiful and gawkily elegant, as Ryan cuts the boat's engine.

As we drift, he explains that the posts are not just a perch for birds, but also a kind of grave marker for an old bayou camp.

"Locals would come here—*right here*—and fish and trap and hunt and have fish boils and crab boils and shrimp boils and they would walk out their back door and hunt ducks. And now look—there's not a blade of grass for miles."

It's true; we might as well be in the middle of a lake. In every direction we can see places where land used to be and where we now see only clouds reflected in the water. In the distance small strands of marsh islands barely keep their heads above the tide, just the hair of their grasses showing. In spots we passed earlier you could see dead trees going under.

"This is not something that is happening over centuries," Ryan says. "Just a few years ago I could look as far as I could see and there was grass. Now it's all underwater. Whatever the reason—sea level rise from warming, the land sinking due in part to oil extraction—it really doesn't matter. The point is that it's happening."

I knew the seas were rising, of course, but before I came to Louisiana I didn't know that the seafloor was sinking through a process called subsidence. Over the centuries, sediment dumped by the Mississippi has weighed down the Gulf floor, causing it to literally sink. And as the land sinks and waters rise, saltwater invades the marsh, killing cypresses and other plants that help stitch the wetlands together. Louisiana's erosion rate is the worst in the country and the equivalent of sixty football fields of wetlands are lost every single day. Which means that if you stand in one place long enough, it might just turn from land to water.[15]

Among the things killing the wetlands are straight lines. Nature, of course, isn't very fond of straight lines, and for centuries creeks wound sinuously through this area. But humans long ago decided that winding was not a good way to travel. They dredged straight canals to replace the creeks without considering the consequences. Straight lines are also required for the ten thousand miles or so of pipeline that travels through the wetlands, carrying oil from the offshore rigs to shore, unintentionally ushering saltwater deeper into the marsh.[16]

This morning, looking down from the copter, I could see how these straight lines crosscut the marsh, and I could also see the rectangular holes of water where oil rigs used to be. The juxtaposition of wild marsh and planned grid made me think back to being a kid at the beach. I loved playing on the small sandbar islands that revealed themselves at low tide: when the tide started to come back in, I would aid the rising waters by digging lines across the sandbars with my heel, creating canals for the incoming tide to run through. I would dig a dozen of these lines across the sandbar islands, flooding them before their time. The same thing is going on here on an enormous scale.

⁓

Ryan has stopped the boat for a reason. Though two of us aboard are professors who lecture for a living, we will learn now that we've got nothing on the boat's captain. Ryan, it turns out, is not just a man with energy and passion. He is also a man with a cause.

As we bob on the water Ryan talks movingly about the loss of people's livelihoods, the loss of animal habitat, the loss of human culture that has accompanied the disappearance of the marshes that made this one of the most biologically productive estuaries in the world. The barrier islands and outer marshes have always been the frontline of defense against hurricanes, and now they are the frontline against the oil, too, keeping it from working its way into the heart of the wetlands. A second defense, Ryan explains, is the Mississippi River itself, which had done more than its part to keep the oil at bay, its massive outward flow pushing back against the Gulf's inward surge. He worries that this might change once the river's seasonal strength wanes.

"The river protects these marshes," he tells us. "But it's also what made Louisiana. The sediment it brought here, the nutrients that helped grow these wetlands."

And the river would still be doing this if it were not hemmed in by the levees.

"What we have to do is redistribute," he says.

He doesn't mean the wealth, God no. He means the water. "Free the Mississippi," it turns out, is Ryan's rallying cry. He is not talking about radical freedom here, since without the levee his lodge would be underwater; what he is really looking for is a series of diversions so that the river can feed the marsh at various points, rather than dump all it has to offer in one great slug out in the Gulf.

"You know, it's funny," he says. "A little while ago I was in Alabama on Lake George with some friends, and they said, 'Oh, I wish those boats wouldn't go so close to the shore or they'll cause erosion.' And right then it dawned on me what the rest of the country thinks erosion is: a little bit of dirt falling down a bank. But when we speak of erosion down here we are speaking of millions of acres of land going away, never to return. And the only thing that is going to make this land come back is the same thing that built it in the first place. The Mississippi River. All we have to do is let the river go through these marshes like it did for eons of time when it built Louisiana. We have to break it out of the levee and reintroduce the river through different diversions and spillways. We could start slow, maybe one diversion channel, but that would be sufficient to bring in the freshwater and to grow the freshwater aquatics and to keep the saltwater at bay and start to rebuild Louisiana. If we let the freshwater start flowing into the wetlands it would start growing the land that very first day."

I think of how the river looked this morning from above: corralled by its levee, segregated from the wetlands. In Ryan's vision the river would spread out more naturally, like a watery hand, feeding the marshes with nutrients it has gathered during its powerful crawl and sludge from Minnesota down through the country's middle and finally to the Gulf. Of course, far from "natural," this would be a massive engineering project on the scale of building the levee itself. But it would be engineering toward a different end, toward releasing the river, to an extent, and letting it do what it once did naturally.

"It is such a beautiful solution and it doesn't just solve the problem of erosion," Ryan continues. "It protects us from hurricanes, and oil, and it tackles the problem of the dead zone in the Gulf. Right now we have a dead zone the size of New Jersey out in the Gulf, where the Mississippi dumps all the crap from a thousand farms—the manure and fertilizers and insecticides—along with the nutrients. This creates algae blooms and removes the oxygen and kills all sea life too. But if this same nutrient- and fertilizer-thick water runs into the marshes, the result is completely different. Everybody says, 'We got to stop the nutrients; we got to stop the fertilizers,' but you know, we really don't. All the wetland plant life will use the nutrients, filter the leftover fertilizer, and when it comes out the other end it will be pristine, crystal-clear water. If we let the river go where it's supposed to go, we will be using those nutrients while also cleaning up the dead zone. Let nature do that herself, the way she intended. We think we're smarter than Mother Nature, but we're not. We can sometimes outsmart her for a lifetime or two but she's coming to get us eventually, and she's coming back to haunt us right now."

I think of an interview I heard with a New Orleans scientist. The reporter kept talking about the oil—the action, the adventure, the disaster!—but the scientist insisted on talking about the Mississippi, which he did until the reporter finally got fed up and ended the interview. A lot of people from around the country are mystified when Louisianans, upon being asked about the oil, start talking about erosion, the Mississippi Delta, the river.

"That's because the loss of the wetlands connects to so many things," Ryan explains when I bring this up. "People talk about greenhouse gases and global warming. But think about what losing these wetlands means. These marshes are like prairies, so rich in grasses, and they produce so much oxygen, you can almost see it pulsing off the marsh. I imagine it shimmering off in waves, the way the heat from a fire does.

"Think about what it means to lose a million acres of habitat like this. How many trees are equal to a million acres of grasslands? Many, many . . . a tropical forest's worth. Then think of the way a healthy marsh reflects back sunlight with its pretty blue water and grasses. What do you think all this black water around us is doing? It's sucking in the heat."

I'm not sure if this is what the Ocean Doctor has come to hear, but either way he's not complaining. I, for one, am spellbound. It is a bravura performance, obviously practiced but also passionate, delivered from the soapbox of the seat behind the boat's console.

"Everybody wants instant gratification; humans only think of their own lifetime. But what happens is while we're thinking in our seventy years, everybody wants their project started tomorrow and then they want it done the next year. It's not going to happen like that: it took eons of time to build this land and it will take time to build it back. But if we don't start right now, my great-grandchild will never see what I've seen and what my ancestors saw. And this part of Louisiana will not be here in thirty years. This is a national treasure, but we're letting it slip right through our fingers. It makes me sick."

Ryan is not a big gesticulator. For all his intensity, he keeps relatively still, hands on the steering wheel. No doubt he has spent a career telling fish stories from that very spot, regaling his customers. He is part nature boy, part show-man, part arm-twister, and we, floating in the middle of the bay, are a captive audience. I'm not sure if Brian and Nathan, drifting fifty feet away in the Zodiac, understand what is going on—engine trouble?—but maybe they have stayed in Ryan's lodge long enough to get it. Whatever the case, they wait patiently.

"I used to see deer and bear and bobcat out here when I was hunting and fishing," he continues. "Now I see rac-coons and otters clinging to little spits of grass that aren't big enough to sustain life. A whole world is going away in front of our eyes. Not too long ago people made their living trapping down here. But there are no more animals to trap. They're dead, there's no habitat. So instead of yelling and screaming because someone was trapping animals, why aren't people yelling and screaming because the animals are dying because there's no habitat. There was once a way of life, but that way of life is gone. People used to hunt ducks for a living and sell them on the market. Well, now we have processed ducks—that way of life gone too. If it keeps on going like it's going there will be no shrimp. And then, what next? This is the best place in the world. And for me not to know that my kids can come and see it? 'Cause it won't be here? Scary."

Ryan starts the boat up again. He seems to be done for the moment. The rest of us look at each other, stunned, and resist our instinct to applaud.

I love Ryan's description of the way energy shimmers off the marsh in waves. And I also love the way it shimmers off the most motivated and driven people. I am energized by obsessed people like Ryan, who manage to unite a wild personal energy—an energy beyond reason—with a love of what they have found here on earth. Running into someone like Ryan is reassuring in the face of a larger hopelessness; it's good to know that if we are going down, at least we'll go down fighting.

Ryan is *greedy* for this wild place, he wants it for himself and for future generations. He *needs* it. Our best hope lies in working with nature, just as we must work with human nature, and that does not mean sitting in a field and picking daisies. It does not mean denying self-interest either. Self-interest, rather than an evil, contains as much energy as anything else on earth. What is a more glorious fuel, capable of getting more done?

I think back to a lunch I had a couple of years ago with Jim Gordon, the president of Cape Wind, who had fought for almost a decade to put a wind farm out in Nantucket Sound.[17] When he first made the proposal I reacted with outrage, like so many other Cape Codders did. "It *can't* happen here, not

in this beautiful place." But I evolved, and that day over lunch
Jim pulled out his iPod and showed me that, though it felt
calm, the winds on the Sound were blowing strong enough
to provide us with around 67 percent of our electrical needs,
even during the crowded summer.[18]

"The environment is changing with or without Cape
Wind," he said. "This region is one of the most susceptible to
sea level rise. Already you've got insurance companies pulling out from houses within a half mile of shore. You've got
more intense storms, beaches eroding. And as the population doubles, where is our energy going to come from? It
would be nice if it were a choice between Cape Wind and
nothing. But it isn't. It's either gas and coal or us. We need to
make some hard choices."

I liked what Jim was getting at, but what I liked even
more was that he admitted his motives were not pure:

"My opponents say, 'He just wants to make money.' And
I *do* want to make money. I want to show that it's not just
coal-driven power or oil-powered power plants that make
money. Alternative energy can make money."

I have held on to that conversation during these dark
times in the Gulf. While this past April was a bad month
environmentally, there was one bright note that did not get
much attention outside of the Northeast. During the very
same week that the oil started gushing, Jim's project, the first
offshore wind farm in the country's history, was approved
by the Department of the Interior.[19]

Perhaps I can better explain what I am talking about by
using an example of what I am *not* talking about. Not long
ago I watched a film of a lecture recommended by a friend
who knew I was going to the Gulf. In the lecture, Jeremy
Jackson, a famous coral reef ecologist, described the current
and future state of our ocean. The news was bleak: corals are

gone, fish are gone, algae blooms are everywhere, and the ocean floors now look paved, all previous growth dug up by trawling that kills the very grounds where future fish will be born. In twenty years we will have only minnows left, and that if we are lucky. Jackson's talk was an apocalyptic tour de force and you could see people in the audience nodding even as their hearts and hopes sank. Then, after delivering his funeral oration for twenty minutes or so, he concluded: "The thing we really need to fix is ourselves. It's not about the fish, it's not about the pollution, it's not about the climate change. It's about us, and our greed, and our need for growth. . . ."

It sounded familiar: we need to change something basic about ourselves. I think Jackson is probably right about the fate of the oceans. Certainly I would not debate him on a subject that he has spent his life studying. But I think he is dead wrong about human nature. I would argue that while he was busy staring down at sea urchins through his microscope, he did not keep quite as careful note of the species he is part of.

"We humans are an elsewhere," said my old friend and mentor, the poet and essayist Reg Saner.[20] The natural human state is that of hunger. We are always reaching, reaching, grasping, wanting to be somewhere other than where we are. It is not my role to stand apart from this and say, "No, it is bad to reach and grasp." That is as foolish as it is ineffective. A better question is how to use this desire, and the unimaginable energy it unleashes. Is it possible to change the objects we grasp for? To refine and revise what we mean by "more" and "better"?

Talk of our doom is supposed to motivate us to change, but most often it leaves us feeling impotent. Rather than cause us to fight, it makes us withdraw. And to set the problem in terms of changing our basic nature is to insure it is a fight we will lose. It would be like saying to a bee, "You'll be okay as long as you stop buzzing and working so hard on the hive." If we set ourselves against human nature we propose an impossibility, insuring our own failure.

The question is not "How do we change human nature?" That has never been the question. The question is "How do we *use* human nature?"—just as surely as it is "How do we use the river or the tides or the winds?" The environmentalism that makes me most uneasy is a rationalist's environmentalism, one that seems to hint at the perfectibility of man. I do not believe that humans are perfectible, or even very rational. We are a tribe with restless minds. We move and we shake and we need fuel to do it. For most of us, there is no greater punishment than sitting still and, faced with our current crises, we are not going to suddenly turn ourselves into Zen monks. Instead, maybe, at best, we can take some of this restlessness and energy and put it to better uses. Maybe we can nudge it in new directions, or, better yet, divert it toward older, deeper channels where it used to run. Maybe, as we do this, we can be guided, not just by the desire for ease, but also by older ideals of sacrifice; of good work and growth and wildness beyond an engineer's dream of straight lines.

Which does not mean we should deny our engineers entirely, just suggest that they work with the world and not against it. We all have an engineer's voice inside us—calm, rational, logical. We need those voices in difficult times, but we have made the mistake of thinking that that one voice is all. It is not: along with that voice we need one that is wild, inspired, simultaneously guided by, but somehow beyond, mere reason. It isn't that I don't believe in reason, willpower, all that; it's just that I believe in this other thing too. And that other thing is where we merge with the world beyond us, a world that does not believe in straight lines. This is not a New Age sentiment. It is rather a very old one, one we need to get back to. The trouble is that we seem hell-bent on destroying the only thing that might hold a clue to an alternative way to be. And that thing we are destroying is a machine of such complexity that it makes our strongest computers look like children's toys.

Over the last few years we have lost a clear-cut definition of what it means to be environmental, and that is good. So many things are mixed up that we have now entered a world where developers can be the good guys. I like that things have become muddied and complicated. I like that, at the moment, two of my favorite environmentalists are a businessman from Boston and a conservative Louisiana fishing guide, both driven as much by self-interest as their desire to save the world. Maybe they are the poster children for a new environmentalism, a hard-nosed environmentalism that sees how wind and water can coincide with profit.

Maybe it is time for the word *environmentalism* to go away altogether. Maybe the word needs to be knocked over and shattered. Whatever we call the shards that are left, it is not time to think in terms of black and white, good and bad. Black and white is what led to the checkerboard grid that covers the delta, slicing and sinking the marsh. What we need is creative, energetic thinking, but thinking that really takes the world into account—what my father called "the real world," though his meaning was the opposite of mine. The real world is the one that has been here for millennia, not the industrial model that has been stamped upon us over the last hundred years.

We need to unleash our imaginations, wedding them to good science and engineering, while working *with* the world.

This is not a conservative or liberal issue. It is a practical one. How will we next fuel our tribe? What juice will make us go? Do we keep pumping what is basically a dry hole, in the meantime taking risks that will destroy not just ecosystems and habitats and animals, but lifestyles and human culture? Do we do this in the name of sucking the last drops out of an old well, an old way? Or do we start doing now what we will have to do soon enough: looking for a new one?

RUSSIAN DOLLS

We reach the outer edges of the Barataria Bay, our boats landing on a barrier island facing the Gulf. It is hot, midday hot, and we climb out and start walking across the island, a thin patch of sand that stretches out for a couple of miles. Until a few moments ago the place was mostly populated by terns. True, there are a few crabs and laughing gulls and millions of crustaceans and insects and about a hundred million microscopic creatures that I can't see, but until five minutes ago, no human beings. Now there are six of us marching across the island, and we do so in a remarkably self-conscious manner, a manner perhaps unique to our species. While terns are poseurs of a sort, aggressively defensive birds that always make far too big a show of defending their turf, they have nothing on humans. We, as a species, may be overly proud of our uniqueness, but one way that we are undeniably unique is in our obsessive need to record everything that we do.

I am the last to leave the boat, which means I get to look at all the others walking in front of me. Here is what I see. Twenty feet ahead walks Alan Guggenheim, who is filming the people who walk in front of him, a group that includes Brian, the Cousteau team cameraman, who is in turn filming the cameraman from the *NBC Nightly News* with the fancy accent, who is in turn filming the Ocean Doctor, who is holding out a microphone and recording (for his radio show) what Ryan, the only local, is saying. Then, what the hell, just to add another Russian-nesting-doll layer to the whole thing, I take out my microcassette tape recorder and speak the words you are now reading.

My head is spinning—this is what we've come to, the crazy self-consciousness of so many recorders, including myself. *Enough,* says a voice in my brain. I decide I need a little break, and veer off from the line of recorders and yell a too-quiet good-bye. I hike off to the east, cutting across the hump of the island. Once I can no longer see the rest of the group I pull out a beer from my pack. It's warm but I drink it quickly anyway, trying to smash the hall of mirrors in my brain. I am Bruce Lee at the end of *Enter the Dragon.*

It is important for me to be honest about my motives. The Vessels of Opportunity captains aren't the only ones who see potential gain in disaster. Writers and reporters and filmmakers are washing up on these shores like so many tarballs. This, after all, is a big story, and these days our story-tellers migrate immediately toward any story deemed big. I am part of that national migration, and I make no claims to be above the baser motives that drive such a movement. But while I want to be honest, I don't want to denigrate my motives. It is easy to take a single swipe and say that the media is bad and superficial, and that all storytellers are just in it for their own sake. Accepting the fact that self-interest drives us all does not necessarily mean racing toward bleak conclusions. To try to make a narrative out of a thing is not an entirely dishonorable endeavor.

One thing I do know: down here the stories gush like oil. By the time I arrived in Louisiana, I thought I had seen enough already, and in fact considered the possibility that I didn't even have to visit Louisiana to be able to tell the story of the Gulf. I was dead wrong. With apologies to Florida and Alabama, this place is so terrifically strange that it makes me want to cry and laugh, usually both within every ten minutes or so. Never before have I experienced so intensely the disparity between hearing a story on the ground, from the people it is happening to, and the way it is told to the country. Never before have I been so deeply a part of the sheer Lewis Carroll strangeness of the modern storytelling machine.

I continue hiking up the beach, which I soon discover is full of small, orange, oil-covered rocks, perhaps toxic but also flat and good for skipping. The tar is thick here and I pick up a handful of clams and find that they are covered in the same orange goo. We are really out on the edge of it now, beyond the millions of acres of wetlands, facing the Gulf, the oil, and the rig. Despite all the self-consciousness swirling nearby, it only takes about a minute or two of strolling along the water for that good beach-walking feeling to come over me. I leave my sneakers above the wrack line and start swinging my arms. This island is a lot like Masonboro, the undeveloped barrier island I often kayak to near my home in North Carolina. Both will be underwater soon, if the scientists are right, but for the moment I'm not interested in thoughts of doom. I'm interested in the heat rising off the sand and the swaying grasses and the hundreds of terns. I stare toward the east, where thunderclouds are building, the usual dark afternoon gathering.

What is real? What is authentic? What stories do we choose to tell? Is it all contrived, with so many of us on this planet pointing cameras at the rest of us? In an age where everything is recorded, how do we wedge downward and find the story we consider true? We, or at least I, need to break away from it all every now and then, but even those breakaways are often later transmuted into stories.

When I curl around the island's bend and can no longer see the others, I feel the old excitement of solitary freedom. I reach some marsh muck where fiddler crabs seethe along the sand, the whole place breathing. The crabs scurry down into their now oily holes, holes that aerate the sands, which in turn encourage plants to grow. Hundreds of these creatures, thousands perhaps, run from me, a giant in their midst. The land hisses with them.

I am briefly free of other people, or partly free, and exhilarated for it, until two sights lasso me back to community and narrative. First is some tampon-boom tossed up onshore, coiled there like a python. My exhilaration is immediately dampened: there is no breaking away from what's going on here. The second sight is more dramatic: a dead pelican sprawled on the shore. Its face is buried in the sand, its wings spread like a pterodactyl, posing in death like an emblem on a flag.

I look at the dead pelican and see no obvious signs of oil. I reach down and unbury its head. Pelicans are enormous birds, mostly mute, and they, like ospreys, have come back from the scourge of DDT to become a virtual living symbol of the Gulf. They serve as Louisiana's state bird and anyone who has spent time on these shores has had the pleasure of watching them dive. The birds bank and twist and plummet as they plunge into the surf, a few of them turning in the air in a way that looks like pure showing off, but all of them following their divining rod bills toward the water. When they are all diving at once there is something symphonic about the way they hit the water, one bird after another: *thwuck, thwuck, thwuck.* At the last second before contact they become feathery arrows, thrusting their legs and wings backward and flattening their gular pouches. They are not tidy like terns and have no concern for the Olympian aesthetics of a small splash, hitting the surface with what looks like something close to recklessness. As soon as they strike the water, instinct triggers the opening of their huge pouch, and it umbrellas out, usually catching fish—plural. While still underwater they turn again, often 180 degrees, so that they emerge facing into the wind for takeoff. And when they pop back up, barely a second later, they almost instantly assume a sitting posture on the water, once again bobbing peacefully as they drain their pouches. It's a little like watching a man serve a tennis ball who then, after his follow through, hops immediately into a La-Z-Boy.

This pelican's diving days are over. I study it for a while, taking field notes in my journal. I think of other dead or dying birds I have seen on beaches over the years, each deserving of a hero's burial. I remember watching a laughing gull die on the shore in North Carolina. I could do nothing to save it but I moved it up out of the tide line so its last breaths weren't wet and gurgling ones. Death is an everyday occurrence on the beach, of course, but for some reason that one bird reminded me of my father when he was dying, and before I could stop myself I was filled with a deep sadness. Its black eyes blinked quickly and wind ruffled its feathers. For most onlookers, this sight, this moment, meant nothing; for that gull it meant the end of the world. An hour later I walked back out and it was still there, struggling, trying to climb to its feet. I remembered my father, referring to himself, saying near the end that he was "hard to kill." The gull, too, like a weed, was not going to go easy. Still blinking, its crimson bill tip dug into the sand, the wind still ruffling the gray primaries. It was a tragic end befitting Lear. Night was coming on and it would definitely not last until morning.

Since animals can't narrate, few of us hear their stories. But here is what the bird's struggle said to me: "I can't stand to leave this place. This place, this world, has been my whole life." To my mind the gull merited an elegy and a requiem.

I feel for this pelican, too, dead during this summer of oil. But somewhere inside me, right next to my more noble impulses, are superficial ones. I would like to keep walking down the beach, would like to continue breaking away from the group. And I should, I should. But this pelican is not just a dead fellow being or an object for contemplation. It is also a coveted prize. Before I know what I am doing I'm returning to the group. I will be the conquering hero; I have found gold for the others. Soon I have gathered up the rest of the gang and brought them back to the spot where the dead pelican lies. Within minutes the bird is experiencing a glory in death it never found in life. Its posthumous media moment includes having TV cameras pointed at it and having its picture immediately posted on Facebook. Though the bird can't know it, there is even an outside chance that it will make the *NBC Nightly News*. Or at least it will if the cameraman has anything to do with it. You can see he is excited now. Thanks to the constant intrusion of his camera, I spent the ride over rooting for him to fall in the water. I don't like his way of monopolizing the Ocean Doctor, muscling in close so that Brian can't get any shots without him being in them. I wonder if he will be squeamish around the bird, but then I remember that his last assignment was in Iraq. He is on more intimate terms with death than most of us.

After assessing for a minute, Ryan takes charge of the situation. He is wearing a camouflage hat and a long-sleeved shirt. He picks up the bird and quickly sees what is what. While it would neatly fit our mutual narratives if oil were the culprit, he surmises the real cause of death. As it turns out, it isn't oil but a too-big catfish that killed our pelican. Ryan lifts the huge bird higher with one hand and with the other reaches into the bird's gullet. Then he draws the catfish's enormous skeleton out of the bird's throat and through its gular pouch like a magician pulling out a sword.

Ryan looks at the bird and then at us.

"He bit off more than he could chew," he says. It is both a good line and a pithy, accurate necropsy.

⁓⁓⁓

Who gets to narrate our stories for us? Who lays claim to them? It isn't the main thing that I've been thinking about down here—I am more struck by the people I meet and the unexpected beauty of the place—but it does run like a leitmotif through my travels. You can see the battle over stories in the BP commercials that just started airing: the sincere African American man who claims to be in charge of helping BP make things whole again and claims to care deeply about the Gulf. Meanwhile a counternarrative is told on the local stage: the mayor of Orange Beach says the man isn't telling the truth and that BP won't return his phone calls.

As of this week, 105,000 claims for damages have been submitted and 52,000 payments have been made, totaling 165 million dollars. (Ryan is not one of those lucky claimants.) Cheery ads are running that talk of hard work and a happy future, and the logo of the green and yellow sun shines on. (When I take a closer look at the logo I find myself wondering if its creator was on acid.) Whatever you think of the ecosystems of corporations, you have to admire their ruthless effectiveness. At the moment, the company is spending over five million dollars a week on

advertising—three times the amount they spent during the
same time last year.[21]

At some point the corporate world began to understand
the preeminence of story. But the BP story, and the national
media story, feel nothing like the story down here. The main
thing that does not come through in the newscasts is the
bristling unease, rage, discomfort, and numbed acceptance
of the fact that a multinational corporation is in charge of
everything. But it isn't easy for individuals, however angry,
to reclaim the story, to compete with the national story that
is being told. Yesterday shares of BP rose 5 percent on the
strength of a new narrative: it now looks like the cap over
the well might actually work.[22] Which leads to a new sort of
local frustration. Will everyone be moving on soon? Either
way, it's dumbfounding to watch the media nod and accept
BP's magic trick of dispersants, as if oil out of sight means
no oil at all.

Before we leave the tern island, I find another pelican, and
while there's no catfish clogging its throat, or other obvious
culprits, we can't say for sure that oil did it in. What we can
say, or at least what the Ocean Doctor is saying, it that there
is evidence of oil here, on the smallest of bivalves, the coquina clams that he sifts through his hand. Coquinas are
creatures of the edge, adapted to live in between the tides,
covered and exposed by the waves. They are also an indica-
tor species, warning of danger to an ecosystem, and the indication right now is oil, lots of it. Colorful bands radiate
outward from the center of the inch-long clams. The Ocean
Doctor explains that coquinas are filter feeders, consum-
ing phytoplankton and algae and detritus and in turn being

consumed by fish and shorebirds. He doesn't need to tell us that if oil is in the smallest of creatures it will soon be in the largest, through the trickle-up theory of the food chain.

We hike back across the island to where the boats are anchored and find Nathan has been napping. He sleeps under a pup-tent-like awning set up over the Zodiac. I decide to abandon Ryan's boat and ride in the Zodiac with Brian and Nathan, fulfilling a childhood fantasy by joining the Cousteaus. Soon we are flying down one of the side canals through the delta, trying to keep up with Ryan.

Nathan captains our boat. If the Cousteau group is a kind of Scooby-Doo gang, then Nathan's role is an intriguing one. Like most everybody else, he can scuba dive and take pictures, but he is by far the youngest of the group, having gotten the job straight out of high school; and while he sometimes serves as a gofer, he's also the group mechanic. When we take off I notice that he drives the boat like a madman, cutting though Ryan's wake, the hull slapping up and down on the waves. The pontoon boat is open and I hold on to a rope so I don't get thrown out. I know that one of the pontoons has already been torn and repaired, and perhaps the boat is better suited to the Pacific waters near their Santa Barbara headquarters than to these shallow waters filled with oyster beds.

We follow Ryan through a narrow channel. A bottlenose dolphin also cuts through the opening with no problem but we end up getting stuck. So Brian and I climb over into the water until we manage to push the boat forward off the sandbar. We reach an island that at first glance seems to be deserted. But then we see the makeshift carport jammed with dozens of all-terrain vehicles. We climb out of the boats onto the island to examine the mountainous piles of huge sandbags, each about the size of a small heifer, that

crowd the inlet between this island and the next. It takes
a minute to figure out the purpose of the sandbags, but it
seems that they are here to close the water flow through the
inlet between the islands. While we're not sure of the intent,
we do know how the sandbags got there, remembering the
Black Hawk helicopters we saw this morning. Sandbags
have always played an important role in coastal battles, and
despite their frequent ineffectiveness, they are very important to little boys on a symbolic and narrative level. They say,
"This is war. This is serious. You nonserious people—you
nature folk and sensitive types and you outdated fishermen
and oystermen—don't understand about war." But the sandbags here are almost comically ineffective. The water, being
water, simply flows around them. We watch a new channel
form before our eyes. Just as we are ready to push off again,
twenty men in orange vests emerge from behind a stand of
junipers down island. They are all African American, save
one, and they all carry large transparent plastic garbage
bags that billow in the breeze like parachutes.

We climb back into our boats and leave them to their
work. Once we are offshore a little, Ryan throws in a line
and almost immediately catches a redfish as we watch from
the Zodiac. The Ocean Doctor will take samples from this
fish and the others Ryan catches. These samples, we hope,
can act as a corrective to the overall lack of testing so far. No
one seems to be testing the fish yet, except for a generally
ridiculed method called "the smell test" (which is exactly
what it sounds like), and so we are taking matters into our
own hands.[23]

Ryan throws out an anchor and pulls up some oysters for
the Ocean Doctor to sample. When I look over again the
Ocean Doctor himself is pulling out oysters and his NBC
pilot fish is pointing the camera at him. "Can you do that

again?" the cameraman asks and the Doc, maybe a bit exasperated, dutifully reaches down and "pulls out" the oysters that are already in his hand.

We drift close to their boat but don't tie up. When we get back, the Ocean Doctor will FedEx these samples to a scientist in Maine for analysis. Ryan is interested in getting the results back as quickly as possible; he wants to try and have them in hand by the EPA meeting in Buras on Tuesday. He is sick of being told that the fish can't be tested yet—why the hell not?—or that BP will determine who gets to do the testing.

What he is really sick of, I suspect, is other people telling his story. Screw that. This is his home, not theirs.

"I'm going to bring the results to that meeting and put them in the face of the officials," I hear him say over the waves. "I'm going to call bullshit on them."

BEYOND FLIPPER

Like any sane person, I am fond of dolphins. For the last seven years or so, since I moved south, we have been on neighborly terms. Now I get to spend the afternoon in their company along with Brian and Nathan. It's time to return to the lodge, but we're not getting back anytime soon. Thunderheads gather in the east, over Buras. The storm brings some pretty serious violence: lightning jags across the sky and rain falls like a gray curtain. We'll be out here for a while.

We have a view of the storm but stay relatively dry. The folks in Ryan's boat pass the time fishing and collecting samples. We putter after them, directionless. Until we see the dolphins. The first sign is the usual one: fins lifting out of the water. Suddenly we are surrounded by them. They dive in and out on both sides of the boat, their gray backs glistening.

We float with the dolphins for company for a while and I remember my first New Year's Day in the South, when I kayaked over to Masonboro Island. Escorted by a squad of pelicans, I paddled across the channel, thinking of birds and looking to the sky, until suddenly, something rose out of the water. A dorsal fin. Then three more, close by. I'd like to say that I reacted immediately with sheer delight at the wonder of nature, but that would be a lie. The first moment was panicked, before slow identification of friend, not foe.

Today there's no panic. We know what we see and soon the Zodiac is smack in the middle of a dolphin pod. They swim close to the boat, almost close enough for me to touch them. I can hear the exhalation from their blowholes. It's a deep watery out-breath, like that of a weight lifter before the push (though they are almost simultaneously inhaling). Of course, the Cousteau boys are no strangers to dolphins either, having swum with them on numerous occasions. Nathan eases the throttle forward ever so slightly while Brian sits in the bow and films them. The dolphins circle our boat, curious, and a few do a little tail flapping, but mostly they seem content to just hang out. Brian starts to put his camera away—"In this light they'll just look like gray blobs on a gray ocean"—when a young dolphin catches us by surprise. We hear it on the port side of the boat, breathing like a pervert, that deep exhalation, and when we look over, its snout is practically in the boat.

"Think of what they're breathing in and out," says Brian, and I do. I think of the coquina clams and the way that, back in Fort Pickens, James described how his wife's bathing suit was stained brown by the water. With each breath these huge mammals are taking in oil and possibly dispersants. It's not a thought I want to dwell on.

On some levels my life has been an erratic one: hard years of debt, failure, and frustration. But one thing I am proud of is this: I have always made an effort to get to know my nonhuman neighbors.

Dolphins have been particularly good neighbors. Moving to the island town of Wrightsville Beach, outside of Wilmington, was a little like moving to the set of *Flipper*.

It's true the dolphins were less interactive than on the old TV show, rarely crying out to you in their ratcheting chatter, never imploring you to save a distressed swimmer or put out a boat fire, but you got the feeling that it was only a matter of time. I'd never felt as unsettled as I did during our first fall in the South. My wife and I had a new baby and a new place, and I had a new job. For all the joys of fatherhood, there were times during those early months when I felt like I was living inside a sweaty nightmare.

But dolphins helped. All through our first fall, I would carry my daughter, Hadley, down the beach in a little papoose contraption called a BabyBjörn. We would stop and watch the dolphins as they lifted up into our world before dipping back into theirs. Thinking it was only polite, I started to teach myself all I could about dolphins. Like most people, I knew they used sonar, but what I didn't know could fill a book. What I didn't know was that just as a person experiences his or her life mostly through sight, and a dog through rivers of smell, dolphins experience the sea around them acoustically. They do this through a process called echolocation, which involves emitting between thirty and eight hundred clicks a minute and sending these sounds bouncing off the world around them and then receiving, and analyzing, what echoes back. In this way dolphins sonically understand both where they are and what is around them; they use this system to place themselves and other objects, the bouncing signals providing complex and ever-changing maps of their underwater world.

Looking back now, I see that learning all I could about my new home was metaphorically akin to echolocation. I felt massively out of place in the South and needed to bounce off everything surrounding me before I could call it home. But to suddenly have dolphins in my backyard! Not long after that first New Year's paddle I paid a visit to Ann Pabst, a professor and marine scientist at the school where I taught. She articulated my still vague thoughts. "The fact is that no other large wild animal regularly lives so close to man," she said. "It's like sharing land with a grizzly bear."

Bottlenose dolphins grow to six hundred pounds and over ten feet long, but at the same time I could see them on an almost daily basis breaking out of the water and into our airy world. I quickly learned many other things from Ann. I discovered that these sleek torpedolike mammals are said to have evolved from a shrewlike creature that hung out around the shore until one day taking the plunge. Now, of course, they have left that water-loving rodent far behind. Though dolphins still breathe air, they regularly dive to one hundred meters, taking advantage of another nifty evolutionary adaptation, a collapsible rib cage to prevent the bends. Under normal circumstances they can travel close to twenty miles an hour, but as many boat owners know they have also found ways of going much faster. Bow riding is a favorite dolphin sport, consisting of riding the wake in front of a boat, using the pressure of the waves to surf effortlessly at whatever speed the boat is going. Like any good surfer, they soon grow tired of just straight riding, and engage in rolls and jumps and twists. Dolphin play and dolphin kindness have become clichés, but the animals are more complex than merely kind, and have also been known to tease and torment the fish that they corral, even those they don't intend to eat.

An exciting discovery was recently made by three scientists, one of whom, Laela Sayigh, lived only a twenty-minute paddle away when we moved to our new Southern home. Dr. Sayigh and her colleagues coauthored a paper that concluded that bottlenose dolphins convey "identity information" with "individually distinctive signature whistles." This may not seem particularly eye opening until you stop and think that it's just a fancy, scientific way of saying that they call each other *by name.*[24]

~~~~~

We watch the dolphins sea-serpent in and out. Their bodies are sleek, black-gray rubber balls that gleam with water. They cycle up and down as if moving in a circle. A constant rotation from air to water. From up close you can stare right into their intelligent black eyes—no pupils, all black. The best moment is when a mother and her two young glide close to the boat. The mother swims and the young ones follow.

Despair mixes with delight, as it often does these days. Seeing the dolphins reminds Brian of the plight of their large cousins, the sperm whales, out in the Gulf. The team has been trying to film the whales over the last few weeks, but this is not an easy task given how few are left and how deep they swim.

"They like to hunt and feed along the continental shelf," he says. "They prey on giant squids that live exactly where the oil is. There are only about thirteen hundred of them left here, so the loss of even one whale is crushing."

It is hard to overcome our anthropocentric bias, to understand that animals have complex lives and that the loss of those lives is not simple, cold fact. But in the midst of considering our own plight—even right now in this oily summer—don't we owe it to ourselves to consider theirs? I tell Brian a story that a charter fisherman named Kit told me back on Wrightsville Beach. We were drinking beers at the marina when Kit, a Hemingway look-alike, described watching a dolphin give birth from his boat. The dolphin baby was stillborn but the mother wouldn't accept the loss. She kept nudging the small dead body up toward the surface with her snout. When the baby got to the surface, it would sink back down. Then the mother would once again push it up toward air.

These sentient beings, these *families,* are now swimming through waters contaminated with oil and almost two million gallons of dispersants. A people who rarely have empathy for *Homo sapiens* from other parts of the world, and who only recently released a substantial percentage of its own population from human slavery, may not be expected to have empathy for *Tursiops truncatus.* So how to stretch our minds, how to understand that this is not just one of the greatest environmental disasters in United States history, but in dolphin history? We will not see the body count, of course, since a dead dolphin, like the stillborn baby that Kit saw, sinks to the ocean floor. Out of sight, out of mind.

<p style="text-align:center">~~~~</p>

One of the reasons I felt displaced when I first moved south was that I defined myself as "from New England." Over the last few years I have gradually begun to redefine myself as a creature of the coasts. As I did so, I began to more consciously explore those coasts, from Alaska to the Outer Banks to Nova Scotia, learning their science and lore. Looking back, it was as if I had been training for something, but I hadn't known what. It never occurred to me that what I had been training for was the Gulf.

I've learned a lot about dolphins in my travels. In the current coastal environment—overcrowded, overfished, overcompetitive, over-oiled—dolphins are often pitted against fishermen in the media, fighting for a dwindling stock of fish. But if you spend enough time wandering the coasts, you see that they are really in the same boat.

Last October I continued my coastal exploration by taking a three-day kayak trip from downtown Wilmington, North Carolina, to Little River, South Carolina. The goal of the trip was to encounter dolphins in their home waters and to explore the impact of gill nets, the fishing nets that can stretch out to one hundred feet long and entangle whatever swims into them, dolphins included.

My environmental bias was clear from the start: dolphins good, nets bad. Which for the most part proved true. But the picture grew more complicated on the second day of my trip when my two companions and I paddled into the small fishing village of Varnamtown, which was a little like paddling into the nineteenth century. Enormous, scarred shrimp boats leaned on the single listing dock that made up the town's center, and old men sat on plastic upside-down barrels and talked about—what else?—fish. It was a pristine, Rockwellian scene, but, like so many things about the coast, it was more complicated than it looked. As well as shrimping, the residents of Varnamtown caught fish, prominently a species called spots, by setting gill nets across the mouths of marshes. The problem was that fish weren't the only thing they caught.

I approached the men on the barrels who, despite my accent and funny kayaking clothes, acted relatively friendly. Cautiously, I brought up dolphins, which they, to a man, called porpoises.

At first no one wanted to answer my outsider's question, but after a long pause a guy in an ancient baseball cap took a shot at it.

"No one wants to catch a porpoise," he said slowly. "But sometimes it's hard to catch one thing and not hurt another."

He went on to tell me that Varnamtown had only six hundred residents and that for generations most everyone made their living by fishing. But the monofilament gill nets were not part of that tradition, being a relatively recent innovation first used in the seventies.

"If you want to know more about the nets you should talk to Ed," the guy in the cap continued. "He's made and mended them as long as we've had them."

Ed lived two miles up the road, so I left my friends with the kayaks and set out in my soggy boat shoes. But after a quarter mile, I came upon two men, who both looked like they were in their twenties, and who were lolling on the grass in front of the modest town hall. They asked me what I was doing and, when I told them, the younger one introduced himself as Henry. Henry asked if I wanted to borrow his bike.

"You can just leave it in my backyard when you're done," he said. "My house is the white one by the end of the road near the docks."

Which was how I found myself pedaling down the road to meet Ed, who was soon enough greeting me on his front porch. Ed was a friendly and relatively toothless man in his eighties who showed me his arthritic hands, the knuckles gnarled like knots on a tree branch, by way of explaining why he didn't work on gill nets so much anymore.

"My doctor says I should quit altogether, but what else would I do?" he asked me.

We walked over to the small skiff that sat up on sawhorses on his front lawn and he pulled out a net to demonstrate how it was used. He made a particular point of showing me the breakaway connectors between the strands—they looked like plastic carabiners—and mentioned they were there so that whales could break through the nets if they got caught. When I brought up dolphins he shook his head.

"It's hard to catch a porpoise," he said. "I don't think you have to worry too much about them. Anyways, it's the gill netters, not the porpoises, who will soon be extinct. The spots just ain't running like they used to."

We walked back to his porch where he took a seat in a rocking chair while I sat on a cooler. More than one dog barked inside.

"The fisheries are dying," he said. "And I think pretty soon they'll just go. Gas prices, regulations, less fish. It all adds up."

He slowly shook his head.

"It reminds me of the woman who kept cooking hard pancakes for her husband. She kept cookin' them hard and he had no teeth, like me. He never said anything but one day he went in his bedroom and got his shotgun and shot her. Well, he didn't just shoot her 'cause of that one time. It all piled up."

He was smiling and wore a look that assured me that the anecdote did not come from personal experience. We talked for a while more before saying good-bye. I pedaled back to the docks, but I couldn't find Henry's white house and didn't know where to leave the bike. Finally a neighbor, an older woman, came out and asked me what I was looking for. When I explained, she shook her head slowly and pointed at a house that was pale green, not white.

"He should know the color of his own house," she said. "Since he painted it. " Then, almost under her breath, she muttered, "I hope Henry's not on drugs again."

I laughed when I retold the story to my friends back at the boats, but about half an hour later, paddling downstream from Varnamtown, I thought of what Ed had said. *It all piled up.* What could young Henry do, born of a town and a family that had always fished? Loiter on the town hall lawn? Do drugs?

All around us there were signs of change along the coast. As we paddled, a gang of Jet Skiers came roaring down the waterway like a motorcycle gang, spouting their rooster tails of water. On the banks, trophy homes puffed out their chests, not ten miles downstream from the formerly isolated Varnamtown. Brunswick County was like every other county along that stretch of coast: barely populated twenty years ago, the only towns being clusters of summer beach getaways and old fishing villages. Now, after two decades of Realtors successfully billing it as a retirement community—the new Florida!—it had become a whole different place. During that time over one hundred golf courses were built in Brunswick and neighboring counties. Maybe Henry could grow up to be a golf pro.

Before we left the river we cut across to the small town of Sunset Harbor, where I beached the kayak and went in search of a man the locals called "Yankee Dave." I found him at his fish market and we talked dolphins for a while. He'd moved down from New York over a quarter of a century before but still carried his birthplace in his local nickname. Yankee Dave was an old-school fisherman—he had been gill netting just that morning—with a new-school twist. He not only used the whale-release rings that Ed had showed me, but he checked his nets frequently and had educated himself about dolphins. He told me that just this past weekend he had witnessed a dolphin necropsy at the university performed by my colleague Ann Pabst.

"It was a young male," he said. "It's the young males who usually get themselves tangled up. They don't know the world yet. They are reckless, shot through with testosterone."

It wasn't until much later that it occurred to me that this comment might also have applied to Henry. Ed had managed to make it to the end of his life fishing and making nets.

But the young males of Varnamtown, reckless and full of testosterone, not yet knowing the world, could no longer turn to working the waters the way generations of young men had before them. Nor could they turn to those older men as models. It all piled up. If you grew up in a fishing village where everyone had always fished, what did you do when there was no more fishing?

The Gulf is a land of birds and a land of dolphins in part because it is a land of fish, which both birds and dolphins feed on. As do the humans, by the way, who have drawn not only their food, but their culture and way of life, out of these waters.

Almost all the restaurants I've passed in the last two weeks have been seafood restaurants. Back in Pensacola, Florida, where I first saw the Gulf waters, I drove by a huge sign for Joe Patti's Seafood Company, and, curious, pulled into the lot. What drew me in was not just the sign but the docks on the side of the building. An old blue shrimp boat called the *Miss Ann* listed at the dock, its outrigger arms tucked in like a giant resting butterfly. I immediately thought of Varnamtown.

I walked into a room as big as a basketball court, full of ice and fish. It didn't take long to notice that the fish, like me, came from far away: haddock filet from Gloucester, snapper from the East Coast and the Yucatán, grouper flown in from Costa Rica. The point was obvious enough: *not from the Gulf!* I milled for a while and then talked my way into a meeting with the store's owner, offering up a hardcover of my book about ospreys and the fact, mostly true, that I was only here to write about birds.

Frank Patti shook my hand hello and said the only birds that he knew much about were "water turkeys," by which I assumed he meant cormorants. I also assumed that this bit of ornithological self-deprecation wasn't true, and sure enough he, like most water people, knew his birds.

Frank handed me a business card with a cartoon picture of his father, Joe: thick black glasses resting on his substantial nose, and a Brillo Pad of hair. Frank was a softer version of his dad. Short, wide, and strong, in a jaunty hat and suspenders. He invited me into the inner sanctum of his office where he no doubt often held court, having his employees as a captive audience. His deep, drawling voice was both his strength and weapon. It was beautiful, with weight to it—full of rumble and growl, rich and resonant. Frank told me that his parents started the business and sold fish out of the front of the house where he and his six siblings grew up. "You got to know your fish in that house." Fishing and fish, he said, were his whole life.

"Fishermen are like farmers," he said. "But they're farmers who can't see their crops. It's never been an easy job. Now it's impossible."

There were those rushing to proclaim that Gulf seafood was safe, but he knew it wouldn't do to sell it again until they were sure it was clean. He would rather delay the reopening of Gulf waters, knowing that opening too soon, and selling tainted fish, would be the worst result of all. Seafood, after all, was all about trust, and trust had been lost. His had been a thirteen-million-dollar-a-year business, but profits had been cut by more than half. The chemicals, in particular, eroded trust.

"The oil is bad enough. At least with the oil, Mother Nature would have a shot at cleaning the Gulf with the winter storms. But the chemicals are worse. Nobody knows what they're going to do. They sink to the bottom and that's no good. The food chain starts at the bottom."

I asked about the docks next to the store and he told me that he had two boats he used to get all of his shrimp, one that he owned and one owned by a Vietnamese captain. He had more than a grudging admiration for the Vietnamese.

"They resurrect old boats. They go out in boats that no Americans would go out in."

Now both his boat and the Vietnamese boat were out of commission, or rather had been recommissioned. No one in the world wanted Gulf shrimp right now and so the two had become Vessels of Opportunity. The pay was pretty good— two thousand dollars a day for his boat and fifteen hundred dollars for the smaller vessel owned by the Vietnamese. But Frank was skeptical.

"What's going to happen once they plug that hole? Is the money going to dry up too?"

It was then that Alice Guy, the store manager, chimed in. She had been working at the computer off to the side of the room, but it turned out she had some strong ideas about the spill. To educate herself she had been reading Charles Wohlforth's *The Fate of Nature,* a book partly about the long-term consequences of the Exxon Valdez spill. "It lets me know what's going to happen here eventually." She had long dark hair and a not-quite-thick Carolina accent, and she pointed out that as soon as you signed on with BP, you also signed away the right to criticize your new boss. She had a very clear opinion about what it meant for BP to pay the fishermen.

"It's just hush money, plain and simple," she said.

On the other hand, who could blame a fisherman for signing on with BP? Was there any other choice? No one had fished in these local waters since May. No one had legally caught an oyster or a shrimp.

Looking back on that meeting, and on my trip to Varnamtown, I find myself wondering about the fate of all fishermen in the coming years. What will happen to the Henrys of southern Louisiana? Like most young people, those here model themselves not just on their elders but on the way their elders live, a way that has been slowly extinguished and that, this summer, has been put completely on hold. At this moment no one is fishing, no one is shrimping, no one is oystering. Sure, the fishing grounds will reopen, but when? It is hard not to feel we are at the end of something.

<center>~~~~~</center>

Brian, Nathan, and I watch the dolphins for a while longer, though they seem to be getting bored with us. They swim farther away from our boat, but before they exit, they provide one final treat. A large individual circles back and slaps its tail, a big sharp crack. Then it dives down and swims off. I wonder out loud if the dolphin we saw was "fishwhacking," which consists of batting fish with their tail flukes so that the poor stunned fish will sometimes fly thirty feet in the air. The truth is the dolphin might have slapped its tail for any number of reasons, including sheer exhilaration. Who knows? Dolphins are generalists, with no one set of fishing behavior, and adapt differently in different places to the local tides, geography, and fish population. In other words, they are creative thinkers, not inclined to doing things just one way.

Ryan signals from the other boat that it's safe to head back. He pushes his throttle forward and starts flying toward Buras. We follow. I hold on to the bowline and stand up as if surfing, bouncing along, wondering if any bow riders will join us. I love the feeling of being out here, of salt and sand and sun, of having spent a day outdoors with dolphins, of living out a childhood fantasy by riding in a Cousteau boat. But the feeling is not a childhood feeling. I know too much. I can't stop thinking about the dolphins. I can't stop imagining these smart, interactive, family-oriented animals swimming through an oil-and-chemical pool of slime.

Is there anything good in all this ugliness? I have heard a hundred people say that perhaps it will lead to a time of reckoning. That even those of us who would rather not think about these things might find ourselves thinking: just what have we done?

But these questions will fade when the spotlight does. We humans can handle only so much guilt, and we grow weary of the work of empathy. Soon enough the national media will skip on to its next big story.

The problem is that there will be many, both dolphin and human, who won't be able to move on. This is where they are from. And they will stay stuck here in this place they call home. Mired.

It is the locals who often take the hit in service of our global needs. Dolphins are locals here, more local even than the Cajuns and their drowned camps, and they have their own local culture that will be lost along with the human one. From different places spring different dolphin qualities. When I paddled from North to South Carolina, for instance, I stopped at Jeremy Creek in the town of McClellanville, South Carolina. There the local dolphins are famous for self-stranding on the town boat ramp and eating fish out of people's hands. In other words, this is a tradition particular to the place, taught from parents to children. I only bring this up so that we don't fool ourselves into thinking it is "just animals" we are killing. These are beings with cultures and traditions.

When we tally up our ledger sheet of gains and losses we had better consider this. We are gaining oil, yes. But one of the things we are ripping apart is culture. Just as Ryan won't be able to show this land to his grandchildren, so too dolphin communities may not be able to continue living in a place they have lived for centuries if we keep taking wild risks for the last drops of oil. The *Deepwater* spill might not do them in, but what about the next one? This is part of the math, too. When we consider what we are sacrificing at oil's altar we had better not forget certain mammalian neighbors, neighbors who live in communities, mourn for their dead and stillborn, teach their children, and call their friends by name.

My initial reaction to seeing dolphins near my home was an almost aesthetic one, like *ooh*ing and *ahh*ing at a particularly pleasing painting. But dolphins are not paintings and they are not symbols of my attempts to find a home. They are the true locals, and I've come to believe that it's just common courtesy, no more than good manners, to treat them with respect. This is not "environmentalism." It's looking out for one's neighbors.

A NIGHT AT THE FISH CAMP

Back at the lodge we all gather around the big screen for a modern ritual. The nightly news begins. I have not watched it once since coming to the Gulf and I find myself laughing at the way the newscasters tell the story of this place. It is all about capping the oil, told in the bold strokes of an action film. Or the newscasters are comedians telling this story broadly, a kind of somber slapstick. Whatever it is, it has nothing to do with the place or the people I'm meeting.

The coverage might be shallow, but there is no lack of dramatic tension in the room. The suspense revolves around whether the adventure of finding our dead pelican, already beamed to New York by the NBC cameraman, will make the cut of tonight's news. Hope starts to fade, however, when our trip doesn't make the first two segments. A truck commercial comes on—"Built Ford tough!"—and as the news returns we have the sense that it is now or never. When the next feature is about Chelsea Clinton's wedding you can feel the air go out of the room. Everyone jokes around but there is real disappointment. Was our trip worth it if it didn't make the news? Does a modern story exist if it isn't televised?

I get over the disappointment soon enough. I've got a big night ahead. After the news, I call Anthony to apologize for being late, but he tells me to take my time. I don't want to keep him waiting much longer, but this packed day demands at least a brief respite. All of us, sun- and windburnt, head out to the little deck in front of the lodge. We rest in the shade while Nathan looks over my car. The car has been acting up, and the engine light is now flashing red. Worse than that, a noise like shattering chandeliers emanates from under the frame. When Nathan comes back to the porch, wiping his hands with a rag, he declares it fit to drive, at least up to the Myrtle Grove Marina where I'll park for the night. As for the shattering-glass music, his advice is to ignore it.

It would be fun to sit here and drink a beer with my new friends, but I have places to be. Right before I say good-bye, David Guggenheim emerges from the lodge. He has been on the phone with the scientist from Maine who will analyze the samples we took today. The Maine scientist says she doesn't need the whole fish, and before the Ocean Doctor hurries off to FedEx the samples, he holds out his hand to show us what he will be sending north: the small purple liver of a redfish.

As I drive up to the Myrtle Grove Marina my car bucks. When I pull in for gas I am approached from the other self-serve pump by a guy who looks suspiciously like the serial murderer in *No Country for Old Men*. He asks, and so I tell him, about the problem with my car.

"There's a lot of water in the gas down here," he says.

With that irony in mind, I head up to the marina where I will leave my troubled car for the night. I grab my backpack and a cooler of beer, and go in search of Anthony. Soon enough I find him, and he is not alone. He had mentioned on the phone that he would be accompanied by his uncle, a somewhat wild-eyed man who, I am guessing, is already a bit inebriated. I'm not one to throw stones on that account and we shake hands before climbing into their boat and pushing off. For a young kid, Anthony handles the boat like a pro. I'm not surprised to learn that he grew up out here on the water.

It's dusk, a beautiful night. The marsh brims with insects and birds. It doesn't take a lot of imagination to guess why Anthony's uncle is coming along. I can just imagine what the boy's parents thought about the prospect of him spending a night alone in a fish shack with a middle-aged Northerner. Probably something close to how I felt when he told me over the phone earlier that Uncle Lyn was coming. Visions of *Deliverance* danced in my head. But those visions flow both ways I suppose.

If Lyn is acting as a chaperone, he is clearly a buddy, too. And—I am guessing here—he may also be a kind of alternative role model, playing against what I imagine to be Anthony's more strait-laced parents. Whatever the case, it turns out that this is not a rare trip for these two. When they head out to camp they are Huck and Jim: they fish together, hunt together, get away from civilization together. And they tease each other, too. Anthony's angle of attack seems to be his uncle's incompetence, but it only makes Lyn laugh. At the end of almost every sentence they call each other "Bert," a kind of manly pet name. Halfway out of the canal, the older man cracks a Miller Lite, which I take as a signal to reach into my own cooler and respond in kind. We toast. Though we come from different parts of the world, we speak the universal language of beer.

The early moments are spent sussing each other out. I'm not quite dorky enough to try and sprinkle in a "Bert." But maybe I try a little too hard to be hardy. They regard me with mild amusement and a little suspicion.

One thing, though, becomes clear pretty fast. They are

happy that I provided an excuse to get out here. They *love* their fish camp.

"This is the best place," Anthony says in his rich, gurgly drawl as we pull up to the dock.

It's hard to argue. Around the rickety shack the bayou is alive to the point of overdone, full of ducks and swallows and herons.

"Lookit that," Anthony says suddenly.

I think he is pointing at the egrets that perch in the cypress about a hundred yards on the other side of the marsh—a fellow bird-watcher!—or maybe at the fish jumping below them. But then I see a line in the water heading directly from our side of the marsh toward the jumping fish, and then the line seems to crumple the water a bit while also rising slightly out of it.

"A gator going after those redfish for dinner," he says.

We watch the gator for a while and then Anthony and Lyn start unpacking the boat. I offer to help but they insist that I sit still, and so after a quick tour of the spartan quarters—a combo living room/kitchen and a back bunk room packed with single beds and a bunch of pictures of Anthony and his family all either on the water or in it—I sit with my beer on the lopsided dock and watch the gator and the birds and the now-dying sun.

A few summers ago I visited Kerry Emanuel, an MIT professor and one of the country's leading authorities on the recent intensification of storms. After our interview ended, Emanuel said something that I have not been able to forget. As I was walking out, I pointed to a Hopper painting of a house by the sea and we moved from talking about storms to talking about homes along the coasts. Emanuel described what he called the historical "natural human ecology" of the coast.

"The natural human ecology of the coastlines tends to be a few castles or mansions built very solidly that will withstand anything nature has to throw. But only a few— everything else is sea shanties. Which the normal person would just go to for a weekend. These shanties or cottages are disposable and people don't put anything of value in them and don't insure them. Every now and then they get wiped out and that's to be expected. It's the same all over the world . . . it's very democratic."

In other words, through most of human history people *expected* their coastal homes to be occasionally destroyed. This made sense to me, and still does. To live by the water is a joy but part of the contract is the acceptance of the ocean's caprice and violence. The acceptance that nothing is permanent. Now as I examine my surroundings, I see that Anthony's fish camp, for one, has no pretensions to being anything other than a shanty. Everything about the dock and shack is askew and plywood-patched, barely looking like it will hold together. It is a house that seems to float, not just on the water, but on a sea of grass.

Since this place was underwater during Katrina I assume it has been rebuilt since then, and when Anthony comes back outside, I ask him about it.

"Somehow we only lost the roof," he says. "We don't really understand how."

So his was a shack that was *not* destroyed. We talk some more and it is clear that even though Anthony comes here a lot, he is excited to be back. "People love their camps," said Leona, the woman at the marina who helped put me in touch with Anthony. "It's where they really live."

"It's a place to get away from it all," Anthony says, but

then I notice that he has a pretty funny, and contemporary, manner of getting away. As soon as he has unpacked, he cranks up the generator and blasts country music on the stereo and turns on the old TV in the corner of the make-shift living area. The first lyrics from the stereo I can make out are, "We just met and I don't know if we'll make it / But you look good in your shirt." Once all the appliances, including the generator-powered AC for the bunk room, are blaring, Anthony calls his parents on his cell phone to report in about his strange Northern guest. He tries not to let me hear, but I catch the gist: I am okay, and it looks like I won't try to kill or rob or rape them. The feeling is more or less mutual.

I drink a couple of beers with Uncle Lyn and actually do feel like I'm getting away from it all for a minute. Anthony strolls out and we talk a bit. It turns out that after school he sells boom, of all things, and did so even before the disaster. I make a lame joke about it being a "boom time" for his business, and they understandably don't crack smiles. I ask Anthony what he hopes to do after he graduates from high school. I would say "when you grow up" but the fact is that he seems pretty much grown, and remarkably composed for a sixteen-year-old. He begins to talk and I learn that, like Mark Twain before him, his ambitions are to be a boat pilot on the Mississippi. I mention that just this morning I saw Pilottown from the air and he seems impressed. Pilottown, after all, is the hoped-for destination for all who share Anthony's ambition. Pilottown is his Paris.

Having stated his dream, Anthony becomes taciturn, perhaps too embarrassed to elaborate. It occurs to me that down here saying you want to be a riverboat captain might be a little like saying you want to be a starting point guard in the NBA. Only a very few will make it. While I stare out at the darkening sky, Anthony heads to the upper deck to help Lyn get dinner together. They are cooking something in tinfoil on a small, portable charcoal grill. A few minutes later, Anthony brings me a plate of rice and something white and boney.

"It's redfish," he says. "We just caught it this morning."

Caught it here, it goes without saying, in the poisoned Gulf.

It doesn't matter. I take a bite and it is good. It might be dangerous to eat the local fish, but I am so fully in when-in-Rome mode that I would probably dig heartily into a baked nutria rat if that was what was served up.

As we finish up our dinner, it occurs to me that this is the second redfish I've encountered today. I think of the Ocean Doctor and the sample he took. Mine is a whole new way to take a sample and I wonder if I should put my liver on ice and FedEx it to Maine.

It might be a mistake—it is a mistake of course—but as the night wears on, and the insects drive us inside, I start to talk politics with Anthony and Lyn. They are pissed about the way things are being run down here but also pissed about Obama's oil moratorium. It turns out that Anthony's mom is in the oil business. She is not alone: in the Gulf over 100,000 people are employed by Big Oil. On the other hand, over 500,000 people have jobs in tourism, the industry that oil has now effectively shut down.[25]

Add all these things up and the sum is despair. Without fish and without tourists and without oil what do they have down here? Then add in the fact that most people feel all but ignored by those running the cleanup.

"It isn't just the fish that lost their habitat," Anthony says. "They forget all about the little man."

That of course is the other gulf. The great gaping space between the corporations and the people who have lived and fished and hunted on this watery land for so many generations. *They forget about the little man.* I think of something that BP's chairman Carl-Henric Svanberg said a couple of weeks ago. With a hurt look, he told the West Wing press corps:

"I hear comments sometimes that large oil companies are greedy companies, or don't care, but that is not the case in BP. We care about the small people."[26]

The small people?

We talk for a while more. We bring up that much-bandied-about word *freedom,* and the lack of it, the sense that most people here have that, on top of the deep depression that comes with loss of livelihood, there is also now a crippling sense of servitude. Servitude toward one's country, toward the mission of cleaning up these beautiful and abundant waters, would be one thing, Lyn and I agree. But the serflike sense that one is enslaved to the very master who spoiled those waters in the first place is another entirely. You can almost see the green and yellow sun branded on their skin.

Even more than between Ryan Lambert and me, there is a gap between me and my hosts. But if we are virtual clichés of the Red and the Blue, at the moment we have more in common than usual. I am growing increasingly angered with the way the president has handed off all responsibility for the spill to a corporation. Despite his recent stabs at populism, Obama feels like he is far away from this place, removed both emotionally and physically. As for my hosts, they would continue to blame the president if he threw on a scuba outfit and swam down to the cap to plug all the leaks himself. But to some degree I share their anger and frustration.

What ends up saving us, and keeping the conversation from descending into argument, is football. The TV has been on as a backdrop but when a picture of Drew Brees flashes on the screen they erupt in appreciative whoops. This leads to our mutual enmity toward Peyton Manning, me as a Patriots fan and Anthony and Lyn as passionate followers of the Saints, who bested Manning, a former hometown hero, in the last Super Bowl. I had always thought that the Patriots' Super Bowl win, after 9/11, was one of the most emotional ever, but as they talk I understand that it has nothing on what the Saints' win means to this tattered and water-ravaged land.

Football proves a soothing drug. We talk for a while, and they make a list of their all-time favorite Saints. I get anxious when the first few players they mention are white, but relax when they include Sam Mills, the noble, razor-smart, and undersized African American linebacker who died young. For me it's been a long day, and I thank them for their hospitality and retire to the bunk room. I like the fact that a day that began with my flying over these wetlands will end up with my sleeping right on them. The AC is blasting and the room is dark and cold and my *Deliverance* fears have fled and I am unconscious almost as soon as I hit the bed.

⁓

One thought that has been plaguing me as I travel is that I am living wrong. That maybe we all are living wrong. I don't want to push the thought too far, to get too moralistic about it in the usual environmental way and use it as a club, but there it sits. It's not just all the driving that we're doing—everyone barreling about in a mad rush. It's more than that. It's the sheer speed and constant distraction of most of our lives. "Death by a thousand cuts," a friend calls it. It's as if we were all being jolted with a cattle prod every few seconds, just to make sure we don't concentrate on any one thing for too long.

By contrast, I remember spending one summer as a teenager working on a tuna boat. For the most part the work was boring. But not the nights, when we would putter back into the harbor and I would climb up into the crow's nest and watch the sun set after a long day out on the water. I would find myself thinking that this was not such a bad way to be. How to live authentically in this fractured, virtual world? It is a question of no small concern to me since I suspect I will only be alive this once. To those anticipating multiple lives or looking forward to happy times in heaven, perhaps the issue is slightly less urgent. But I—between replying to all my e-mails and checking my cell phone and crossing items off my ever-longer things-to-do list and putting in the day's requisite mileage—hope to spend a few hours each day doing what I have, in my better hours, deemed most important: making art, being with my family, and traipsing about in what is left of the so-called natural world. To make your one life on earth align with the ideas you have about this life: is that so crazy? To make something other than money. I am not suggesting that we all throw up our arms and eat acorns and wander around all day. But we have set up a world that demands a frenetic, nearly desperate pace that pokes at our attempts at longer attention like stabbing shards of glass.

Time outside works as an antidote to that fractured way of being, and I have come to believe that when we divorce nature from our lives we suffer in ways our brains do not fully understand. In Fort Pickens, I watched an older woman dive into the water for her morning dip. It seemed her daily ritual and she was not giving it up. There are those who don't believe in the oil, as if it were a fairy tale or superstition. Maybe she was one of those. Or maybe she had just decided to stick to her old ways, despite the new crisis.

There is good reason not to want to believe in the oil. A whole way of life, a watery life of not just fishing and shrimping but of beach-going, has been uprooted. I think again of the ospreys that filled that park. Both training from their parents and instinct compel the birds to dive down into the oily waters, and they are not about to suddenly stop doing something their species has done for thousands of years. The lives of *all* Gulf fishermen have been shaken by the spill, and ospreys are the purest sort of fisherman. But unlike human fisherman, they cannot suddenly transform themselves into oil spotters and skimmers.

Maybe fishing people are more like fishing birds than we may first realize. Maybe the fishing humans will find their current adaptation is not sustainable and they will long for their old ways. Some of us, after all, can't abandon older ways that easily, no matter how poorly suited we are to these harried, jangling times. "It's not just the fish that lost their habitat," were Anthony's words last night.

There it is exactly. The corporate model is bad enough by itself, but worse is that it's a bullying way of being that insists on stamping out all other ways. Not just "I want to do things my way," but "*You* have to do things my way, too." Which is to say this is not just a suicidal lifestyle but a murderous one. Because we choose to live this way, because we sign on with it, consciously or not, we tell others—whether they are dolphins, ospreys, or fishermen—that they can't live the way they have always lived. We eliminate their possibilities

along with our own. We cut them off from tradition, the past.
And we cut them off from the one thing—nature—that offers hope of a real alternative.

There's a story you hear a lot down here. You hear it from
locals and you read it in the paper and I think it was even
printed the other day in *USA Today*. It is a new story, but
it has taken on some of the resonance of myth. It will be,
when all this is over—if you believe in such a thing—one of
the legends of the spill. The story goes like this:

He was a charter fisherman, a good man, who led what
the writer Henry Beston once called "an elemental life," a
life of the outdoors, of wind, sun, rain, sweat. Of course
he lost his job when the oil started pouring into the Gulf,
followed by the chemicals that were poured on the oil. Of
course he had little money saved to support himself and his
two daughters and his wife. And of course he hated working
for the idiots at BP—they didn't even know how to tie a line
to a cleat, let alone a single thing about the Gulf weather—
though he admitted that the overall purpose of his new job
was not a bad one: to spot and get rid of the accursed oil.

But then came the fucking form. It was over fifty pages
long, filled with a hundred questions that had nothing to do
with running a boat. They wanted him to fill it out before
sending him out on the water, *needed* him to fill it out, they
said, in able to properly insure themselves and compensate
him. But it made doing his taxes look like a joke—and if
you're self-employed, taxes are no piece of cake. He resisted
filling it out and grew angrier. A lot of things were pushing him. The oil spilling, the BP folks running the show, the
prospect of a future where the Gulf yielded no fish or, if

it did, no one wanted to eat them. But he focused all of it on this thing, this *form*. It came to symbolize all that was wrong with what was happening. What is a form after all? A shape, a physical embodiment. And this thing, this form, became an embodiment of everything that was wrong in his world. What had been right in the world? Even before the oil there had been a ton of stress, and the fish weren't as plentiful as they had been. But still there were pleasures. The camaraderie of the docks, the satisfaction of pulling out in early morning, a fresh start each day, the cyclical feel of both the days and the seasons, the fact that he knew this place and that on the best days he puttered back in as the orange ball sizzled down into the water and, despite all the crap, what he was doing felt right.

The form was not about the season's cycles and was not about hard, honest work. It came from the top down the way the whole disaster did. Of course, he was angry and depressed about what was happening and it would be wrong to say that the form, a mere stapled sheaf of papers, was the reason he killed himself on his boat. But it would not be wrong to say that those stapled pages symbolized something to him, something that he was perhaps unable to put into words, and it would not be wrong to say that it was in those pages that that something—a feeling of rage, impotence, and sorrow—found its lasting and final form.

~~~~~

I have not been able to sleep much down here. Or rather I sleep short and hard. I've always gotten up early, but these days I am combining that with going to bed late.

Today I wake in a black cave filled with stereophonic snoring. Anthony and his uncle sleep on the bunks on the opposite wall. I grab my things and stumble out to the living room, where the TV is still on, the emergency signal wailing. I look at the clock: 4:30. I want to go back into the cold cave, close my eyes, and forget about the world, but I know that won't happen. Instead I make a large pot of coffee and face the coming day.

I head out to the front dock with my coffee cup and take a seat. I scribble down notes from yesterday, trying to remember everything that has happened before it slips into the past. I wait for the sun. There is enough of a porch light for me to write, but the canal is dark. Even the birds are quiet, except for the occasional croaking of a heron.

I've been up and writing for about an hour when I hear a motorboat puttering down the canal toward me. An early-bird fisherman no doubt, so I keep my head down in my notebook, ignoring the boat. At least until I hear a voice come out of the darkness.

"Morning, David," the voice says.

Until this second I felt impossibly remote and removed from the world. I wouldn't have been more surprised if someone had said "good morning" through the flaps of my tent while I was solo camping in Antarctica. But if this place is the end of the earth, it is also, for the moment, the center of it. I get up and walk down to the edge of the dock and the Zodiac clarifies itself in the light and I recognize Holly as the person who said good morning. Then I see Brian and Nathan and the rest of the Cousteau crew, heading out for a day of looking for oiled birds.

They pull up to the dock and we laugh at the strangeness of our encounter. After a minute they push off and no sooner have they puttered away than I find myself wishing I'd climbed into their boat and headed out with them.

With the sunrise my regret fades. As an amateur naturalist,

I couldn't ask for a better spot. Barn swallows swoop below the eaves behind the house where an oyster midden slopes into the water. A green heron with a mussed-up haircut hunts patiently on the back dock, surprised and irritated to have early company in its daily spot. The heron waits above the water on the dock, peering down, until it's time to stab a fish. And it is just one of one hundred morning hunters. The next two hours are a festival of bird life: various bitterns and cranes and herons swooping this way and that. Obviously I will never be as attached to this place as Anthony is. But I, too, love it here. Love the surroundings and love that this shack doesn't pretend to be anything more than a shack, making no claims to permanence.

Nature was our first home, our old home, and, to paraphrase Emerson, we miss it dearly. I am not saying that we should all run off and find cabins in the woods. There are no more cabins anyway. No places apart. Think of this place, this fish camp, seemingly remote, but vulnerable to the tendrils of oil. I'm not talking about "getting away from it all," but its opposite: acknowledging where we came from. How to really understand that this thing we seem so dead set on destroying is our home and that we are—still—a part of the world we grew out of? I'm not suggesting we need to have a perfect relationship with so-called nature; that we need to grow zucchinis and wear flowers in our hair. But if we don't need a pure relationship we do need *some* relationship.

That is one advantage of a shack like this. Nature is not some box over there. It is the world that has sustained us and given us our connection to any sort of elemental, not merely virtual, life. When we divorce nature from our lives we suffer in ways our brains don't understand. To walk by the shore, to swim in the sea, to fish, and feel the sun. Could it be that we are willing to give this up for the comfort of forms and straight lines? It's as if our new credo were, "This thing, this business model, developed over the last hundred years or so, this system that gives great rewards to a few *Homo sapiens,* is superior to the vast and complex machine of life of all beings that has evolved over billions of years." Do we really believe this? Could we? Maybe the answer is "yes." Maybe we hate uncertainty so much, and are so intent on stamping it out, that we don't mind also crushing the living world in the process. Maybe our twin gods of ease and speed have ascended above all else.

One thing I like about a shack like this is that it's honest. It admits that the world is uncertain and impermanent and that the ground is never firm, that sands shift and islands migrate. For most of us the fact that this same world is wild, joyous, dramatic, and enlivening somehow does not make up for its messiness. We are quick to sell our birthright if things are convenient and quick and straight. Sacrifice is an outdated virtue. Better a controlled castle than a shack that can be wiped out at any moment.

~~~~~

As you may have gathered, I am feeling exalted as the sun rises. But I soon learn that it is the dawn rush hour not just for the birds but also for the Vessels of Opportunity. Here they come, down the canal, and soon I'm witnessing an aquatic traffic jam. Boat after boat flies up the canal then slows down in front of the shack, respecting the No Wake sign, before speeding up again. The Vessels aren't the only part of this new ecosystem: the laughing gulls, searching for scraps, follow the line of boats as they head out to sea. I see a boat full of African American men, snug in their life vests, and a single white man who is steering the boat. This is a common enough sight on these cleanup crews.

As the sun rises in the sky the bird action slows and the Vessels come fewer and farther between and still my hosts sleep on. I am bored and, when I notice one of the VOOs coming back in, I yell to them. They ease up to the dock and I ask if I can get a lift into the marina. It's probably against some rule but the guy captaining the boat says yes anyway. I run in and grab my bag and leave an envelope of bills to pay for my stay, and scribble a quick note for Anthony. Then I jump on the boat. I am ready to go, but first the captain hands me a life preserver. "It's the rules," he says. I put it on and, feeling snugly encumbered, take a seat in front of the console as we head back in to the marina.

III. "Firm Ground Is Not Available Ground"

BARRIERS

I am exhausted. I want to sleep but there is the little matter of where I'll be spending the next night. At first I imagined I was only going to stay in Buras for a couple of days, but there's no way I can leave this place just yet. It feels, as I said, like the center of the universe. On the other hand, I can no longer impose on Ryan's hospitality by staying at the lodge. Ryan and Lupe have been more than generous, putting food in my belly and a roof over my head for drastically reduced rates, but enough is enough. Before I headed out to the shack last night, I packed everything in my car and am not entirely sure where to go now.

But a plan is forming. The environmental magazine I'm writing an article for is part of an environmental organization that has a headquarters in town, just around the corner from Ryan's lodge. There they have set up a kind of base camp for those reporting for them on the spill. When I first traveled down here it looked like I would be able to sleep there, but the organization's lawyers decided that, while I was under contract and could work there, I was not officially an employee and so couldn't stay overnight. Two days ago, when I first got here, Rocky, the manager of the environmental headquarters, mentioned he was heading back home to see his family for a week. He told me that the lawyers still had not cleared me for staying nights at the headquarters. But he also told me that no one else would be around and that I would have the place, with its phones and computers, to myself. And then he did something foolish. He handed me the keys.

And so I drive my bucking car back to the headquarters and throw my backpack into one of the empty, unused bedrooms. It's a one-story shell of a building that was likely rebuilt after Katrina. Across the street stands an old store with shattered windows and the words "Celebrate the Birth of Christ" scripted in broken lights on the storefront. From here it's a five-minute drive to Ryan's and a five-minute walk to the Mississippi. In short, it's just where I need to be. I check my e-mail, call my wife, and make elaborate plans for the rest of the day. But before I can proceed with those plans I lie down on my new bed. Just for a minute. Hours later I wake, feeling groggy and drugged.

I turn on the coffee machine and gather my journals, maps, notebooks, and sketch pads. *Headquarters* is a much less romantic word than *lair*. But this will be both HQ and lair for the next few days, and I have everything I need here, including books, coffee, cigars, beer, and computers—not to mention a huge map of the Mississippi Delta on the wall. In short, I have everything required to try to start making sense of the whirl I've found myself inside.

I'm ready to work, to start putting the last couple of days on paper, but there is a small problem: my stomach growls. I decide to walk over to Ryan's lodge. Lupe is there and she insists that I eat a bowl of beef stew. I don't want to be a freeloader, I try to explain in broken Spanish—what is the word for "mooch"?—but on the other hand it smells very good. I'm in the middle of my second bowl when Ryan walks in.

"Who says there's no such thing as a free lunch?" he bellows.

He is smiling when he says it. Lupe hands him a bowl and

he plops down next to me at the long camp table. Before long we have wound our way back to his favorite subject.

"It isn't going to be easy to free up that river," he says. "We have so many forces working against us. The Army Corps of Engineers is a major stumbling block and the navigation people, you know they're fighting it; and the oystermen are fighting because they worry about too much freshwater but they're coming around because they know if it's not done we *will* lose Louisiana. You know? When somebody starts chopping down trees they worry about the spotted owl and the whole world gets up in arms. This place should be like that. Right now the Gulf is a secret. It shouldn't be a secret. The best thing about the spill would be if it made people think about the Gulf. If it brought this place to light. Everybody in the world ought to know about it, and come in here and help me change it. We've got to think creatively. We can take it back but we're going to have to get the Army Corp of Engineers out of our way, and start opening up these spillways and let the freshwater go."

He points out the window, across the river.

"Now all the old cypress trees are dead and gone. There's stumps left but when I started guiding there were standing trees. All gone. And where you can see this plain as the nose on your face is where that levee is. You look on that side of the river, without the levee, and it's solid beautiful live oaks, but you look on this side, the levee side, and it's solid death. Stumps. Because we've shut the river off. It's so plain to see. Why doesn't anybody do anything about it? I need help. I need the environmental groups to get onboard and come force the Corps of Engineers to reintroduce the Mississippi River to these marshes. We can't wait any longer. It's going away more rapidly than I've ever seen."

After lunch, back at headquarters, I consider what Ryan has told me. For starters, I understand his antipathy toward the Corps of Engineers. The Corps are the kings of straight-line thinkers, applying the geometric mindset of engineers to the messy, fluid, and often very unstraight natural world. They have done more to alter the shoreline of this country than any other entity, excluding Mother Nature herself.

I have gotten to know the Corps by traveling the coasts with Orrin Pilkey, a professor emeritus at Duke University and one of the country's leading experts on coastal geology. For decades Orrin has warned against overbuilding on our shores and in the meantime has become a galvanizing figure in the coastal battles around the country. This has made him many friends and many enemies. "An idiot with a beard," a North Carolina politician recently called him. But there are those who see him as a hero, fighting against overdevelopment.

Orrin is a great connector of things: of the disastrous consequences of having over 50 percent of our country's population living on a narrow fringe of coast that constitutes only 17 percent of our landmass, just as the shoreline is eroding and the sea level is rising and coastal storms—including, most obviously, hurricanes—are becoming more violent.[27] Well before others did, Orrin saw this combination of forces as the recipe for disaster that it is. His longtime rallying cry has been "Retreat," advocating that we move back from the shore. But Americans don't like the sound of the word, especially Americans who have invested millions of dollars in their homes. And to the Corps the idea of retreating from the shore sounds positively unmanly.

Theirs is a simple enough philosophy: *Defend against and fight the sea at all costs!* To do so they erect walls; build barriers; throw up sandbags—anything to stop erosion. The only problem with this, albeit a small problem from the Corps's point of view, is that it doesn't work. In North Carolina sandbags cover vast stretches of shore, though to call them "bags" is to not get the point across. They are enormous, ten feet long and terrifically ugly, great lumpish loaves that transform beaches into war zones. On an island to the north of my home, the residents have become so habituated to the sandbags that last summer a friend of mine saw a young woman in a bikini sunbathing on top of one.

The problem with sandbags, Orrin explained to me when we first met, is the same as with seawalls and jetties and groins or any sort of wall: they are there to protect the buildings, not the beach, and ultimately it is a healthy beach that makes a healthy shore. This makes property, not people or habitat, a priority. While the sandbags might temporarily protect a particular building, they simultaneously cause down-drift erosion. This means that on the Atlantic coast the next house down, the house to the south, bears the brunt of the erosion. And so, while piling up sandbags might understandably be viewed as self-protective—*I can't let my house topple into the sea*—the act is also, at the very least, an unknowing assault on one's southern neighbor. If you put up sandbags, the next house to the south will have to put them up and so on and so forth.

If you are willing, for the moment, to imagine Orrin as a sort of coastal superhero, complete with an "R" (for "retreat") on his chest, then his archenemy would be the Corps of Engineers. (They might sport an "S" for "straight lines" on their chests.) They frustrate his attempts at creativity, just as they do Ryan's. At every turn they seem to act in a way that contradicts what Orrin believes. They build groins and walls and pile sand and sandbags to defend coastal homes, not understanding or believing that by doing so they are destroying the very beaches that protect the shore. They can't seem to wrap their heads around the idea that tides do not respect a wall; that the effect of a wall is to destroy the slope of a beach until there is no slope at all. What you are left with is water slamming straight against a barrier without any of the gradual run-up that a beach provides.

Most people, especially newcomers, have a hard time accepting the sheer uncertainty of living on the coast. "Firm ground is not available ground," wrote the poet A. R. Ammons.[28] He was speaking specifically of beach grass, which has trouble finding purchase in the shifting sand, but also of the difficulties that anyone or anything has in living by the sea. To live by the ocean is to know uncertainty, whether you're a mollusk or a lobsterman or a bottlenose dolphin. Here sands shift, islands migrate, storms assault. John Keats wrote of the poet's "Negative Capability," an ability to be in "uncertainties, Mysteries, doubts."[29] It's a quality that comes in handy when living on the shore.

Living on the shore also means having to get to know some basic facts about the place, having to accrue—what exactly?—is *wisdom* too corny a word? Okay, scrap wisdom. But how about something much simpler: watching and learning how the world works and then letting it work with you and not against you? It is what we humans have done for centuries, not because it was sensitive or groovy or environmental, but because it was effective.

~~~~

At the end of my first week in the Gulf, after traveling from the panhandle of Florida west to Alabama, I visited an island that embodied all of the issues Orrin Pilkey has spent a lifetime fighting against. After leaving Fort Pickens, I set up camp in the home of Bethany Kraft in Mobile, Alabama. Bethany was a new friend who I met at a conference where I'd taught earlier in the summer, and also the executive director of the Alabama Coastal Federation. Generous but beleaguered, she told me about her battles with BP as she tried to help clean up the coast. Like my waiter at Applebee's, Bethany believed that everything was connected, a belief that has practical implications in her work along the shore. Which is what had made the cleanup effort so frustrating, as organizations seemed to be operating in separate cells, not communicating with each other, and the BP leaders kept her volunteers off the beach.

One morning Bethany organized a trip down to Grand Bay, a beautiful, sprawling marsh situated along one of the small patches of coast that Alabama has been allotted. Our guide on the marsh was a friend of hers named Bill Finch, a bespectacled, ponytailed naturalist who looked more like he came from Berkeley than Mobile, and who has known and studied the marshes of southern Alabama most of his life. We spent a delightful morning knee-deep in the marsh, with Bill Finch gently lecturing us on "connectivity issues" and telling us the Latin names of insectivorous plants. Later, after sloshing our way out of the marsh, we drove over the long bridge to Dauphin Island, Alabama's one true barrier island, and I watched a transformation come over our guide.

Things were going fairly normally, though we were still wearing our marsh-wet sneakers and pants, when, heading down toward the west end of the island, we hit a roadblock. I felt like we'd driven right into a joke: a cop, a rent-a-trailer, and a woman who looked like my grandmother in a security uniform blocked the road. We asked what was going on. "They're padding the beach," the cop said. None of us knew what this meant despite the fact that our car was loaded with coastal experts. "Padding the beach" was a new one on us.

We tried to talk our way in; Bill, after all, was a well-known local environmental reporter and Bethany the head of an environmental organization, but no go. "You can park and walk to the beach, but no cars," the cops told us.

So Bill, a little too fast, pulled his car into a side road, hit the brakes, parked, slammed the door, and started marching down the road on foot. He walked impossibly fast, and I, after losing time changing from wet sneakers into kayak booties that looked like black ballet slippers, was too far behind to catch up. Bill, it turned out, was pissed about being kept out of a place that he considered part of his home range. In my former life as a cartoonist I would have drawn smoke coming out of his ears.

Beyond security, we found huge piles of sand, some thirty feet high, lining the beach on the south side of the island, and farther to the north two more rows of sand ran in lines down the island's spine. Trucks, their beds spilling over with sand, rumbled up and down the road. Having lived on a barrier island for years, I instantly knew what was up: sand from the calmer northern side of the island was being carted over to the Gulf-facing south side to protect the homes that were threatened by erosion and storms. What I didn't know yet, but would soon learn, was that this was all being done—trucks had been running up and down the island for two months already, and millions upon millions spent—under the auspices of protecting the islanders, and more importantly the homes of the islanders, from oil. In fact the residents had been trying to bolster the sand in front of their Gulf-side homes for years but some very sensible environmental regulations had prevented it. This sort of project is euphemistically called "beach renourishment," though it nourishes nothing. In fact, the reason it was banned was because it destroys beach ecosystems. But when the oil started spilling, these keen-eyed opportunists saw their moment. They petitioned for some of the millions that BP had given the governor of Alabama and, since all rules were off during the gold rush of emergency, they finally got not just the beach renourishment project of their dreams, but had it all paid for. You may think people who take advantage of a disaster are venal, but you can't say they aren't smart.

Bill was so far ahead now, with Bethany a hundred yards behind, that I decided not to bother to even try to catch up. Instead I climbed over one of the sandpiles that looked like they were dumped there to fill a giant's sandbox, and down the other side. Abruptly, I found myself on a small skirt of beach. No wonder those homeowners were desperate for the sand, I thought. From the water there was only about twenty yards of sand and then a sharp scarp, or sand wall, above which teetering trophy homes sat. One house already had waves beating against its foundation.

I am not heartless. I understand why a homeowner would want to try to pile sand in front of their home to keep it from falling into the sea. But there are a couple of problems with this approach, the first being that it is almost always a mere postponement until the next storm drags the new sand away. Millions of dollars are spent and then nature does what it was going to do anyway, just a little later. The larger problem is that this "solution" shows an almost complete misunderstanding of the way that barrier islands work. To start with, *barrier* is really the wrong word; while these islands do defend the mainland to some degree, they are essentially permeable. They survive through a method quite different, and more fluid, than that of the *Homo sapiens* that have claimed them as their homes.

Barrier islands *migrate*. They move, and grow, most often shoreward. This is happening constantly, but particularly during storms. The way an island handles a storm is through a kind of elemental judo, letting the water rush over it, its sands breaking down and reforming, retreating to the marsh on its backside, rebuilding in a new place, giving and taking. An island lets the surge flow through it, breathing with the storm, never foolish enough to imagine it can block it. Think of Muhammad Ali fighting George Foreman, the way Ali hung back on the ropes. In short, the island survives through a primal rope-a-dope, an ancient and time-tested technique.

Since homeowners don't like to be told that their backyards are migrating, they draw lines in the sand. But the ocean doesn't care about lines. The residents on Dauphin believe in straight lines, suburbia, surveys. But no matter how much sand they pile, and how many of BP's millions they use, they can't change the fact that they just happen to have built their homes on a living island.

I hiked up the beach on the seaward side of the huge sandpiles. This beach, I knew, wouldn't be around for long. Pelicans swooped over the water and natural-gas derricks dotted the horizon. Far from "renourishing" this beach, the tons of new sand were burying a world of coquina clams and crabs and thousands of tinier creatures. Sand was simply being piled, without thought to the world already there, and that sand would be blown and washed away. I remembered something Orrin Pilkey once said to me:

"People fear storms but barrier islands need storms to live. Storms are the way the islands migrate and the way they build elevation. If sand is not pushed across an island by storms then the island drowns."

Despite the ultimate ineffectiveness of dumping sand as

a defense, it is done all over the Gulf and Atlantic coasts. But if this is a common practice, what I saw next was not. By hiking back across the island to the sound side I witnessed the true and particular genius in what the residents of Dauphin had done. When I finally caught up to Bill and Bethany, it looked like they were standing next to a series of Olympic-size swimming pools, until the pools gradually revealed themselves for what they were: huge holes, now filled with water, where sand had been dug out to create the piles on the front side. Which meant they served a brilliant double purpose. Not only did they "renourish" the beaches for the front-side houses falling onto the sea, but they returned these houses on the tamer sound side to what they once were: waterfront property.

When the island had naturally migrated toward the mainland, it had left these backside residents high and dry, with landlocked docks that ran out from the backs of their houses, docks that once were on the water but now found only sand. Until two months ago. Now the docks reached the water again, the new Olympic pools connecting them to the sound. Which meant everyone was happy! The islanders, both Gulf and sound side, were happy because they finally got what they wanted. The governor was happy because the islanders might vote for him. And BP was happy because these folks sure weren't going to be complaining about a disaster that they had cannily turned to their advantage.

"Everyone is happy," Bill agreed. "Except the island."

And, of course, Bill himself. One thing he was particularly unhappy about was all the beach grasses and plants, grasses and plants that held the island together, now uprooted and dead in the massive sandpiles on the Gulf side.

But if he was unhappy, he was also energized. Gone was the professorial figure of the morning, dispensing Latin names. Here was a new Bill, charging around, assessing the damage, beginning to make calls to his editor and local politicians, ready to uncover this mess.

On the walk back to the car I slowed to take notes—I scribbled down the words "Organized Chaos," which appeared on a sign on the front door of one of the beach houses—which meant I was left behind again. Bill and Bethany made it to the car long before I did.

I tried to speed up, but it was dangerous. A rush hour of oversized vehicles crowded me off the street. I had grown used to all the dump trucks rumbling past, but now, in the homestretch back to the car, dozens of Humvees flew by, too, coming from God-knows-where. Bill was ready to go, eager to spread the news of what was going on here, so he ignored the cop and the septuagenarian security guard and drove past the barricades to pick me up. This created quite a clamor at the security booth. One of their cars trailed us as we left the island, and when Bill dipped in and out of side streets, they followed. When we finally turned for the bridge Bill waved good-bye in the rearview window.

"It's unbelievable," he said quietly. "They claim they are trying to protect themselves from the oil. But these clowns have done a thousand times more damage to this island than any oil will."

———

I spend the afternoon hunched down over a pile of books and papers in my lair, trying to make sense of what I saw back on Dauphin Island. I know it is connected to the oil spill, and not just in the obvious ways, but it's connected on some level that's still unclear to me.

I agree with Bill Finch's assessment of what the residents

have done to their own island. Though not entirely. As a student of natural history, I'm inclined to admire the wondrous industry of various species, and what could be more wondrous than the sandpiles of Dauphin? They are as impressive, in their own way, as the termite mounds of Africa. To take advantage of a disaster in the way they did is abominable, of course, but also impressive. Raw self-interest, like something out of Shakespeare, is thriving down here. It seeps in like the tide. For all the energy the residents have put into their project—trucks and Humvees roaring all day long—they are only postponing the doom of their houses until the next storm. But, morality aside, they are doing something primal and instinctive: they are protecting their homes. Never mind that in defending their homes, they have destroyed their island. The place throbs with misspent energy.

The day after our trip to Dauphin Island I called Orrin Pilkey. Our connection was terrible—I caught every other word—but I did gather that he was on a train barreling though the mountains in the Canadian Rockies. And he seemed to get the gist of what I was saying.

"I guess I shouldn't be surprised," he said when I told him what I had seen. "Dauphin Island has always been the worst. They fly in the face of every good tenant of coastal management. I guess I shouldn't be surprised, but I still am. It's astounding what they've done."

But is it really so astounding?

When I first met Orrin I asked him the same question that you or anyone else would ask. So, if you don't want people to put up walls to protect their homes, what *do* you want them to do? You can't be suggesting that you want them to do nothing.

"Yes, that's exactly what I'm suggesting," he replied. "While having houses plop into the sea may be personally tragic for a few homeowners, it is overall a beautiful sight. It's actually a fine example of good coastal management. The cost of trying to save a few threatened homes is tremendous, and the environmental cost is even greater. To save these houses you would ultimately have to put up a seawall and as sea level rises and the waves get bigger, you would have to build a bigger seawall. By then the beach would be gone and you would not be saving anything worth saving. All you would be doing is protecting a few houses.

"The time to do something is before these homes are built. We need to finally understand that the coastline is a dynamic entity, always in movement, and we can't expect it to stay still for us. Or for our homes."

Retreat from the beach, Orrin has told people who build on these islands, leave your homes to the sea. You shouldn't have built there in the first place. All true, but how likely is it that the truth will convince anyone?

Is there a greater and more persistent force in the world than self-interest? Unlikely. It would even give the tides a run for their money. You can feel this everywhere in the Gulf right now. Down here you can almost sense it: that deep and instinctive human craving to have and to take and then to hold forever. Bill Finch is right: there are forces afoot here that go deeper than oil, forces that the oil is just a symptom of.

"What's remarkable so far is how hurricanes barely slow down coastal development," Orrin said to me when we first toured the Outer Banks. "I was down in Florida after Hurricane Donna hit in 1960 and people said, 'Well I guess this is the end of the Keys.' Of course it was really just the beginning. They started building even bigger places. When the North Carolina coast started being developed heavily we coastal scientists used to say 'What we need is a big storm.' We figured that people would see what a storm did and heed its warning. But then Hurricane Hugo hit and we learned that people start building again almost as soon as the wind dies down. Hurricanes have actually become giant urban renewal projects. The buildings come back bigger than before. But of course the site they are building on is even more dangerous because the shoreline has retreated landward and the dunes have been damaged. But still they rebuild. It's really a form of societal madness. I can't put it any more strongly."

A form of societal madness. Well, maybe, yes, sure. I agree with Orrin and Bill, of course. But while I admit that there is something idiotic about all the rebuilding, there is also something oddly admirable. Nothing, not even repeated beatings by storms, can squelch our urge to live by the sea. The logical thing would be to not build by the water, but every day, as the water rises, more and more of us rush forward to greet it. Right now, for the first time since colonial days, more of us Americans live near the water than not.[30] Maybe we can't help ourselves. And maybe, as deranged as some of our actions seem, there is something buried deep that is not quite so deranged. Maybe people, even the sneaky islanders of Dauphin, are hoping for a connection to the sand and water. Like children playing in the waves, they want to be close to all that beauty and wildness. And like children they build castles, imagining them to be permanent.

Considering that until recently I lived on a barrier island myself, it's hard for me to be too moralistic about the desire to live by the sea. And I can't help but wonder what the odds are of Orrin actually convincing people to back away from the ocean. Of course it makes sense, but when has *sense* ever been the driving force that guides how people act?

~~~~

At dusk I head to the river. From up on the hump of the levee I look down at the roofs of houses, houses that will be hit with a freight train of water if the levee breaks again. It's funny, but Anthony's shack, exposed out on the bayou, actually has a better chance of survival than these houses do. During Katrina this little artificial valley was slammed by both the river and Gulf, but it was the river that did the real damage.

It will happen again. "What's wrong with protecting ourselves?" people understandably ask. The answer is that of course there's nothing wrong with it, though maybe there is something wrong with building there in the first place, particularly if the place is dependent on artificial barriers. As Orrin says, one problem with false barriers and blockades is that they encourage people to live in places they shouldn't be living.

When I checked my e-mail earlier I got a note from a scientist, who shall go unnamed, who said that with any luck Tropical Storm Bonnie, due to hit these parts in a few days, will head straight for Dauphin island and undo the months of illegal sand redistribution. Though it now looks like Bonnie is losing steam, one decent storm will wipe out all those weeks of carting sand from the island's back to its front. The residents will find that the laws of nature aren't as easy to bend as the laws of man. Despite their insistence on their own permanence, those wannabe castles back on Dauphin will soon reveal themselves as shanties, plopping off into the ocean. And walking by this river, above this impermanent scrap of land, it isn't hard to remember that Buras could disappear, too. Here you have a real sense of the land as fragile, low, and temporary, with a corresponding sense that all of our human constructions, however sturdy, are shanties at heart.

But enough with doom. This place, like so many of our places these days, may soon enough be underwater, but that doesn't make it any less joyful to explore. We had better not be too strict in our judgments since so few places we love are "natural" in any full sense anymore. In fact, Cape Cod, a place I am prone to romanticizing, was transformed from peninsula to island a hundred years or so ago when someone decided to sever it at the shoulder from the mainland. The point is clear enough: at this time in our evolution no place on earth is untouched by human hands, no place pure.

I worry about clinging to too-strict definitions during this mess. While I find myself growing more "environmental" by the day, I have a problem when environmentalism gets too rigid. For instance, while the spill, the great shitting of our national bed, and its aftermath are bad, truly bad and tragic, I am right now walking along the banks of the Mississippi, something I have never done before this week, which feels folkloric, not just because I've read so many books about it but also because these muscular swallows are once again shooting everywhere and because reeds and small trees are growing out of this old half-sunken barge in the shallows and because the trees are buzzing with insects and the sun is beating down and the wind is blowing along the dirty river and fish are jumping out of the water and splashing back down and I, smoking a cigar and drinking a beer, am feeling something like pure contentment, a temporary animal feeling somehow unrelated to the disaster a hundred miles south. Because there have always been disasters and there has always been death and there's always been a dark thing lurking right beside the light. And because even when Whitman was whooping and hollering his way around the country there was a Civil War about to break out and TB killing thousands and God-knows-what else and everybody dying at forty-two. But there was still some joy and there were still some euphoric moments and why does that matter? Because that is the heart of what it means to be environmental, or at least half the heart. Because if you strip the thing of its joy then all there is left is finger wagging, and who wants that? And more importantly there's this: why fight for a world that you don't love?

I hike down to the levee to a little marina and then down a side path through the reeds to the surging Mississippi. Someone has set up a lawn chair where they come to have their nightly drink and watch the river and maybe occasionally throw in a line, and I, hoping the owner doesn't mind, claim his seat and toast the river with my empty bottle. I am exhausted and I need a rest from this place, and, oddly, the place itself is giving it to me. And what does it matter that I, one human being from someplace else, am feeling good for the moment and not thinking about the oil which is, of course, bad, so bad? Not much, I think. Not much, or possibly everything.

STRESS BALLS

In the national news it's time to celebrate: the word this morning is that the oil is being capped. Folks seem somewhat less excited down here however. You can forgive the locals for not popping the champagne just yet. From hard experience, they have learned to be a little dubious about ingenuity, American and otherwise.

There's a TV at the headquarters and last night I turned on the news. The national media tell the Gulf story in simple strokes, as a kind of adventure fit for *Boys' Life*. Will they cap the well? Will they fire the evil BP guy? The spill has spurred the creation of new myths and one of the most retold is The Myth of the Oiled Pelican. Pelicans, particularly the oiled variety, have become the media darlings of the spill, and while gannets and laughing gulls and tricolored herons must bristle with resentment, *Pelecanus occidentalis* has claimed center stage. In my head I have come to call whatever the biggest, latest accepted story is "the oiled pelican." The oiled pelican is anything obvious or anyone who tries to treat anything, in our complex messy world, in a simple, obvious way.

The general consensus down here is that the oil was overreported at first but that now it's being underreported. In other words the latest oiled pelican is that there are no oiled pelicans. It's dumbfounding to watch the media nod and accept this idea when in fact little has changed. On the same day you are staring down at oil on the beaches you can read that there is no oil on the beach. The truth is that the disaster has now exceeded the national attention span and the newspeople are simply sick of it and ready to move on. *Look, there's lots of oil! Oh, now there's not so much oil!* Once the obvious symbols go away—*Look, there aren't so many oiled pelicans anymore*—the media can, too. Okay, back to business everyone. Time to swing the spotlight elsewhere. What have those kooky Tea Partiers been up to anyway?

⁓

I have stocked up on salami, soup, coffee, and beer from the local grocery store, a place with a third world feel, where cans and boxes lie strewn on the floor or, even when the goods are stacked, this is done with no apparent theme. At the moment the headquarters looks similar: the place is strewn with clothes and notes and empty coffee cups and beer bottles. I've made it my own.

I'm up at dawn and typing. I find myself writing the story of my trip even before I've fully lived it, eating my own tail. I spend a few hours in a mad frenzy of narration. One thing I'm doing today is collecting my thoughts about Anthony's fish camp and about castles and shanties in general. There's a secret in there somewhere, an Icarus-type story about how we fuck ourselves over by always striving to be castles instead of accepting our modest shack selves. But for now I don't quite have the nut of it and have to content myself with circling 'round.

I try to pull some loose strands together. Back when I lived on Cape Cod I was friends with the great nature writer John Hay, who was then in his eighties. When he first built his house on Cape Cod, in 1946, he had a recurring dream that the house was floating out on Cape Cod Bay, rolling on the waves. In a fashion, he would later come to see his dream realized in another's home. John and I once made a pilgrimage to the beach in Eastham where Henry Beston, whom many considered to be Hay's literary ancestor, had spent a year in a cabin and written the classic nature book *The Outermost House.*

"I came up here during the great nor'easter of '78," Hay said to me as we stared out at the beach. "I watched as the storm came in and battered Beston's cabin. Then it dragged the house out to sea. I watched it bob away on the waves. The windows looked like eyes."

After that I started collecting other stories of homes with watery foundations. One of the most arresting was that of an acquaintance of mine named Charles, whose artist parents, Bob Thielen and Virginia Berresford, owned a cottage on a beach on Martha's Vineyard. In days long before the Weather Channel, the hurricane of '38 caught Bob and Virginia off guard, and suddenly waves were pouring up over the dunes. The couple and their cook, Josephine Clark, tried to hike to high ground by Tiswell Pond, but winds were blowing close to a hundred miles an hour and the incoming water was up to their knees, then their waists, then their chests. Soon the three had to swim, which Josephine did not know how to do. She drowned just out of Bob's reach.

In the midst of all this horror, Bob looked back to see an awful, marvelous sight. There was their house, torn free of its moorings, coming after them, floating across the rising water like a ship.

Not long ago I got to see a copy of the painting that Thomas Hart Benton made of the scene called *Flight of the Thielens.* The wild swirl of the black water and the unmoored house looked just like I had dreamed they would.

Now I can add Anthony's fish camp to my collection. I scribble down a sketch of it riding out Katrina. How did that shack only lose its roof?

I remember another story I heard a couple of weeks ago when traveling through Bayou La Batre, a fishing village forty miles or so south of Mobile, Alabama. Until this summer, shrimp had been king in Bayou La Batre. During non-oiled years, the town was part of a Gulf shrimping community that provided the United States with three-quarters of its wild shrimp.[31] But in this unusual year, the whole fishery had been shut down. The timing of the spill was woefully bad as the brown shrimp season usually opened in June and one bad season could do in a boat and crew, particularly in a town that had barely recovered from the devastation of Katrina.

The town had a main street, but the bayou was its true thoroughfare. I followed the long dirt road that followed the bayou, past dry-docked shrimp boats and dredge boats and one old trawler with trees—small junipers—that grew right out of its belly. The road stretched on past a shipyard where people were sanding down boats, and I was soon thinking of the marina near where we'd lived on Cape Cod. But at the end of the dirt road I came upon a scene that was new to me, or at least new since landing in the Gulf. Once again I was in a Spielberg film and the alien autopsy was underway. With the fisheries shut down there was really only one legitimate way to make a living around here, and that was by working for BP. Two hundred trucks sat in the lot and people in yellow and orange vests were pointing this way and that. Cops were everywhere, strutting about, while behind them, out on the water, Vessels of Opportunity streamed forth to spot oil and lay boom. Hundreds of mismatched boats, comically close together, headed out to sea, one after another, as if filming a remake of *The Russians Are Coming*.

They were a motley fleet going to war: single-console boats, shrimp trawlers, pleasure boats. The water was orange and stank but that didn't stop a squadron of twenty pelicans from wheeling overhead and a hundred laughing gulls from heckling the outgoing boats. I pulled over and got out of my car and it seemed only a matter of time before someone questioned me. But that didn't happen, my potential interrogators too absorbed in officious bustling and sheer busyness to bother the guy with facial hair who stood in their midst taking notes.

I headed back into town to eat at the Blue Heron Café. My waitress was Miranda, a smiling young woman, happy to chat. She understood that I would not be ordering oysters or shrimp, and assured me that I was not alone in that decision. "Even the locals aren't eating seafood," she whispered. I settled on a roast beef po'boy, which turned out to be the best thing I'd eaten yet on my trip. The gravy was so good—smothered on both the roast beef and the french fries—that I asked her if I could go out to the kitchen to take a picture of it. She said it was okay and I, apologizing to the cooks, wandered back and snapped a shot of the brown goo in a pot on the stove. Before I left, Miranda and I talked a while. She couldn't have been twenty, but she was sharp.

"We've lost a lot of our lunchtime crowd," she said. "Not just because of the oil but because now BP is feeding them."

She explained that BP had set up its own catering for those who went out on the Vessels of Opportunity. Everyone, she said, was taking BP's money. But she had her doubts. What would happen when BP declared the whole thing over and done, and the money went away and the fish were still tainted, or at least regarded as tainted by the rest of the country?

"The solution is temporary," she said. "But the problem is permanent."

Before I left I asked her about Katrina. She told me how the storm had swept through Bayou La Batre, washing away homes. Grand Bay, where I would later walk with Bill Finch, had been twenty feet underwater.

And then she handed me the small gem that I am adding to my drowning house collection this morning. It was the story of her cousin, who lived outside Bayou La Batre on the nearby Heron peninsula, with marsh on all three sides of her house. Her cousin's family had been warned to leave before Katrina, but they didn't. The waters swept through the house and took the family with them. Miranda's cousin had an eighteen-month-old baby and she clasped the baby tight in her arms as she waded across the flooded marsh. Like the Thielens, they first walked, later swam. Unlike the Thielens, no one in the family drowned. Clinging to her baby, the young mother swam and slogged until she reached the first building on high land, the first building above the water. There they climbed a hill onto a small island of dry land and took refuge in the church.

In the afternoon my lair, my fortress of solitude, is invaded. A young woman arrives and takes the room at the other end of the hallway. Her name is Jessica and she's a publicist for the environmental organization.

I confess immediately, telling her that I'm staying here illegally. I explain my logic: I figure that no one else will get in trouble since they explicitly told me I couldn't stay here and therefore they can't be liable. Which means they can just blame me. She nods. She doesn't seem to mind.

I assume she will be creeped out by the disheveled, hairy man who has left beer bottles and coffee cups all over. I am wrong. What really creeps her out is the idea of being down in this house by herself. This is Buras during the spill: I don't even rate on the creepiness scale.

"I'm just glad someone else is around," she says. "It would be uncomfortable to stay here alone."

I've so quickly habituated to this place that I've somehow managed to forget that Buras is, along with being beautiful and fascinating, kind of scary. *Peligroso,* as Lupe put it when I'd asked her about this town.

It turns out that one of the reasons Jessica is here is to attend tonight's EPA meeting in the Buras Auditorium. I was planning on going too, but then heard that the Ocean Doctor hadn't gotten the results of the redfish samples he'd sent to Maine and therefore Ryan would not be able to "call bullshit" on the authorities at the meeting. In fact, Ryan had decided to head home to Lafayette.

So part of me just wants to stay back at headquarters and keep typing, but at the last minute I decide to go. The town hall is only about a hundred yards from the headquarters, and I step out front and join the crowds massing that way. When I walk inside I find hundreds of people, more than I've seen in all my time in Buras, crowding the hall.

I take a quick look around. At Table Three, the Louisiana Spirit Coastal Recovery Counseling Program is handing out blue rubber "stress balls," though I don't see a lot of fishermen squeezing the little toys. If these things really did relieve stress, you'd have a roomful of Captain Queegs, obsessively squeezing and shaking the blue rubber. I grab a ball, occasionally tossing it in the air as I walk around the auditorium, a place that must usually hold high school productions of *The Importance of Being Earnest* but today houses the surgeon general, hundreds of angry fishermen, and half the reporters in the known world.

I am not acting as a reporter tonight but as a naturalist, and, having pocketed my stress ball, I scribble sketches in my journal, noting characteristics in the way of my avocation. You can identify the network reporters by their field marks, even when they are not jamming a microphone in someone's face. They are generally better looking than regular humans, their teeth far straighter than the locals, and some of them seem to speak with vaguely English accents.

Billy Nungesser, the president of Plaquemines Parish, of which Buras is a part, has become a lead character in the spill drama, in part courtesy of much face time with the reporter Anderson Cooper, and he has tried to fashion himself as the voice of the people. He kicks off the meeting with a short introduction. He has a kind of Columbo air, disheveled and occasionally apologetic about his own flaws in a way that really carries some charm and effectiveness, with sleeves rolled up as if to say: "I'm still one of you even if I've been on TV a lot." There is a strategy contained in the brevity of his speech. You get the feeling that if anyone were to talk for very long this place could erupt. Already there is some yelling coming from the back of the auditorium, rabble-rousing cries of dissent led by one ponytailed former charter fisherman who seems to be taking some glee in all the attention. It is mostly the young men who are causing trouble, and I think again about what Yankee Dave said about the young male dolphins being the reckless ones.

Nungesser briskly addresses a couple of the pressing issues. The first is BP's hiring of outsiders, instead of first turning to locals. The second is the word that BP is now making official estimates of how long it will be until things are "back to normal," and beginning to consider offering "full" settlement packages of two years, or perhaps three. These would be one-time payments based on an estimate of how long the fishermen would be out of work. Nungesser assures the crowd that he is on their side and cautions against jumping at these packages, despite the short-term attractiveness of getting a lump of money right now. Then he addresses the breaking news that any work already done for BP will count against this overall compensation. This sparks more catcalls from the back of the audience.

Sweating, gesticulating, apologizing, Nungesser quells what, led by another, might have already turned into a riot. Then he hands things off to the surgeon general, who also plays the "I am one of you" card, which you'd think would be a hard sell for a large African American woman wearing a white naval costume out of Gilbert and Sullivan. But she, too, works the crowd well, and when I hear she is from Bayou La Batre, I understand that she really is part of this crowd. She is unfailingly chipper during her talk and, like Billy Nungesser, keeps it short. After a few minutes Nungesser takes the stage again and disperses the larger group into smaller ones, telling them that each of their "needs"—counseling, complaints about not being chosen for the Vessels of Opportunity program, compensation questions—will be serviced at different foldout tables.

I wander from table to table and then approach the surgeon general. I have never really understood why our chief doctor is a "general" and now I'm wondering, given her getup, if "admiral" wouldn't be more apt. I like her right away, though. She is friendly, jolly even, and has a talent that may or may not be just political—that of lavishing attention on you when you talk to her, or at least appearing to.

Less upbeat is Timmy, the Vietnamese fisherman distraught over the news that BP is already talking about full settlements. He stands with his arms crossed, his face stern and thoughtful.

"What worries me is when they talk about this as being *over*. When they talk about a 'final' settlement package and say they will now estimate how long it will take for the waters and fish to be back to normal. But how can they know that now? How can anyone know? They say they will pay us for two years. But what if it takes five or ten years until people want to buy our fish again?"

I decide to put Timmy's question to Nungesser, who's in the front of the room, leaning against a table, handling questions from a vociferous group of fishermen. There is no real line, just a spread-out gang, and it takes me a while to slip my question in, but when I do Nungesser turns the high beams of his attention on me. He leans closer, looking like an animated and amiable butcher, and, waving his hands to make his points, launches into this answer:

"That's exactly why you don't jump at the packages. We can't have *their* scientists, their people, telling us when it's going to be better. We have to have our own studies, our own scientists, and that will take some time. We can't be rushed into this thing. . . ."

He goes on, sounding pretty reasonable and caring. I have no idea what skeletons are piled up in his closet—all politicians have them, right?—and I know he made some real missteps early on in the crisis, but based on today alone the man passes my own personal bullshit test with flying colors.

As the auditorium gradually empties, I take a seat and sketch those who remain. After a while I call it quits too, and walk over to the Black Velvet Oyster Bar, the only restaurant around, which is packed with people from the meeting. I grab a stool at the bar next to a small intense man with a gold cross dangling from his neck who is staring hard up at the

TV. Soon I am, too, since it proves to be a documentary put out by National Geographic, I think, about the *Deepwater* explosion and spill. Half the bar is watching, in fact, though most of us are straining to read the captions, since the volume is off. What I notice right away is that the language bristles with military phrases—with *attack, charge, war.* In this language the crop duster spraying dispersants is a kind of World War I flying ace, scarf thrown 'round his neck. It's an action film, that's for sure, and we are rooting for the good guys, though few people in this particular room really believe there are any.

My steak comes and between bites I get in a conversation with my intense neighbor. It turns out he is an ex-professional bull rider who now teaches teenage boys to ride, with an emphasis on the Christian aspects of the sport.

"Some people can stay on a bull for eight seconds but not many can do that and make it look pretty. But any time a boy gets on a bull in the first place it shows he is a man. It is a tremendous act of faith and courage. It's more than 90 percent of the people in the world will ever do. My job is to help these kids and if they have a passion and want to do it, to teach them to do it in the safest and best way possible. If they're doing it for the girls, or for other reasons, then they better not do it. They better do it from the heart.

"I'm interested in building character both inside and outside the ring. I've been down some bad roads and I want to make sure these kids do it different. These are some of the best Christian kids I know. Fear is probably the biggest factor in stopping any of us from doing what we want to do. And every time these kids climb on a bull they are fighting fear and showing faith."

I tell him that it doesn't sound so different from teaching writing. But while faith is important in my game, too, I admit to my own relative godlessness.

He pats me on the arm, which for some reason I do not find odd.

"It's okay," he says. "God, like you, came from out of town."

It's pretty cryptic, and I have no idea what he means really, but I scribble the words on my napkin anyway.

It is only when my new friend turns back to the TV that his faith deserts him a little.

"What I seen a couple of years ago was a grandfather and a son and a grandson, going down a boat in the river, going duck hunting. And right off the bat it struck me that this was something the granddad had likely done with his dad and a line probably went back as long as they'd been in this country. And then the line went forward with his son and now his son has a son—the grandson—and is doing the same thing and it's something that they're going to do every year. Until this one. And that way of life could very well end. Now the old line is broken. If we can't keep this stuff out of our marshes we're done."

I've treated this conversation fairly lightly up to this point. It's been another long day and I'm getting weary of profundity. But he's gotten right to the nub of it. Anthony said it: it's not just the fish habitat we're killing here, and Ed, back in Varnamtown, said it, too: it's the gill netters who will soon be extinct. We are witnessing no less than a radical cutting off from our past. An eradication of a way of life that has worked pretty well for human beings for a millennium or two. Right in front of our eyes the line is being cut.

Part of the problem is that we can't go blithely down this road and then decide we want to turn back. That's because we are destroying no less than our own path back. And if that idea doesn't scare you nothing will. We can re-create lots of things, we can even clone people. But what we forget about is that things don't exist outside of their places. A wolf air-lifted into the Rockies can't just relearn a new territory and birds can't be taught migratory routes. It's obvious enough: we grow out of where we are.

I have a moment of clarity at the bar in the Black Velvet. But then the waters quickly muddy again, as they do so often down here. It's never simple. I remember that Anthony was furious with the embargo on drilling, which meant he was supporting the very thing that might destroy the traditional life he loved. And I have no doubt my friend on the stool next to me, for all his earnestness, feels the same way. Meanwhile I, like God, come from out of town. Which means I get to play the outsider, my moralizing freed up by the fact that I get my paycheck elsewhere. We talk a while more and then I finish my steak and say good-bye. I take a walk by the river in the dark. I'm going to miss this place, I think, this green, beautiful, paranoid heart of southeastern Louisiana. I have no authentic connection to Buras, to this part of the country, and when I leave, after visiting New Orleans for a couple of days, I will be heading home to my wife and daughter and we will buy our first home. So I have every reason to get back and I'm anxious to do so. But I feel a strange tug in the other direction, too. I hate to leave this place.

Maybe at this moment in time we all belong to Plaquemines Parish. I can't help but feel that this at least edges close to the truth. That is, I can't help but feel that while most of us may not have any official allegiance to Plaquemines, we are all still a part of this fucked-up place in this fucked-up time.

What I'm seeing down here is the future. For most of us

it may still just be something you glimpse on TV, but this is where we are all heading. Here you can witness the sacrificing of our places, our homes, in a desperate attempt to gulp up what is left. We will do whatever it takes: take the tops off mountains, poison our waters, go miles below the sea. These are the last bites, last slurps, of ice cream, and while they taste especially good, it's almost over. This is an unfortunate thing for our comfort level. But it is not unfortunate in another way. That is because the gift of fuel, which has given us so much, has also smeared a stain across the world.

And here you can feel all that like no place else. This is where it is happening, where you can experience the shuddering of the death pangs. Welcome to the future. While the rest of the country may not know it yet, down here they're already living inside it.

THE OILED PELICAN

My goal before I leave here is ... what exactly? Maybe to start making connections of the sort that my Applebee's waiter suggested. But also the opposite: to see when things *don't* connect.

I think back to my meeting with Jim Gordon, the president of Cape Wind, in Hyannis on Cape Cod. After we ate lunch and drank beers on Main Street, he took me to the beach where his wind turbines would be planted. As we drove I realized, in an Encyclopedia Brown moment, that his windows were up and his air conditioner was on. He turned down a side street and I followed through a crowded neighborhood of working class houses, if you could in fact call them that when they were most likely vacation rentals. Whatever the case, this was not the Kennedys' Cape Cod, and that, I was starting to gather, was exactly the point my tour guide was making. We walked down through a crowded neighborhood to a crowded beach and I began to see that I was, if not being set up, at least being brought to this particular beach for a reason. Whatever else the Cape Wind debate is, it is often portrayed as a kind of class war, with the rich folk opposing the unsightly marring of their views. And clearly that crammed beach, jammed with umbrellas and beach chairs and kids running this way and that, was not a rich folks' beach. In fact we could barely find space to make our way to the water. Once we got to where the waves lapped, we looked out past boys and girls on inflatable rafts and roaring Jet Skis and powerboats. One of the arguments that wind opponents have made is that putting wind turbines out in this water would be like putting them in the Grand Canyon. Jim was using this beach as both prop and stage, and the message was clear: this ain't the Grand Canyon. The question many have asked is, does having a wind farm out on the horizon detract from the elemental experience of the beach? The argument that Jim was making, so far without saying a word, was that this experience was already limited enough, and that the sight of blades blowing in the breeze was not going to detract from it one iota.

He held up his thumb against the horizon. "From here the turbines will be six or seven miles out. They'll be about as big as my thumbnail."

This of course was another big point of contention. How big would they really look from the shore? And what would it mean for Cape Codders to look out at their theoretically wild waters and see what would be, for all its techno grace, an industrial site? While I had deep sympathy with the aesthetic point of view, it was hard to argue that windmills that would appear a few inches tall on the horizon would ruin the place's wildness. "We need to connect the dots," he said to me. That simple statement carried the force of revelation. Connect the dots. Wasn't that the definition of ecology?

"We would barely see the turbines from here but maybe we should see them," he continued. "It's what we can't see that's killing us. Like the particle emissions from the power plant in Sandwich. And the oil being shipped to run that plant."

He shook his head and stared out at the horizon where his windmills would turn.

"Maybe it's not such a bad idea for us to see just where our power is coming from," he said.

It was a pat line, I am sure. He had likely repeated it over and over during the last few years as he fought his campaign to get his project built. And I was not yet entirely convinced, worried as I was about the way that the turbines might interfere with bird migration. Still, at the moment it was all I could do not to holler "Amen!"

We need to connect the dots. It's a sentence that has stayed with me. *I* need to connect the dots. Why? I primarily define myself as an artist and writer; it's my job to create, not make policy. The world has always been in tough shape, right? Doom has always been right around the corner. The smart course for me would be to hole up somewhere, hide out and write novels or memoirs until I die, not get involved in all this crap. There is still beauty, after all, even if, with our shifting baselines, it may soon be the beauty of pigeons and rats. But, no, no, that isn't true. Think of all the beauty I've found down here even during this dark time.

Some of the dots I need to connect are obvious enough: Like the fact that the places we are now getting our fuel from are our most beautiful places. Think of Alaska, or think of the miraculous Mississippi Delta and the Gulf. Another connection is that we couldn't develop these coasts without the fuel that destroys them, and in getting the fuel we destroy the initial attraction of the places. But there is something else we are doing when we destroy so-called nature and that is what I have not quite gotten at, what I am groping toward. . . .

⌇⌇⌇⌇

It's a long drizzly day of drinking coffee and typing, of walking by the river, of finally cleaning up my lair, washing dishes and doing laundry, covering my tracks. It is time to get moving, but I am sluggish. I should be looking forward to eating and drinking in New Orleans but instead I am reluctant to leave Buras. At night I head over to Ryan's lodge only to be disappointed that Ryan still has not returned. Before long, though, the Cousteau gang comes bustling in and I am caught up in the action. "What's going on?" I ask Brian, who has run into the lodge to grab a camera and is now heading back out.

"We found an oiled heron out on the water," he tells me. "We're bringing it to the Fort Jackson rehab and filming the whole thing."

Then he does what he always does: he invites me along. The next thing I know I am following the Cousteau van down the highway to the Fort Jackson animal rehab center. As we drive I learn more specifics. Today, out near the tern island that we traveled to with Ryan, they rescued an oiled tricolored heron, and they will now try to film every stage of that rescue and rehabilitation. We pull into a parking lot and walk through a gate in a barbed-wire fence to a small guardhouse. There we sign in and the guards issue us badges. I call myself a "key grip"—whatever that is—just as Brian told me to. The guard doesn't blink and hands me a white security pass that I hang around my neck. I walk toward the front door, while the rest of the group heads around back, where the oiled animals are admitted. The parking lot overflows with trucks and trailers, some from environmental groups and some from news organizations, and a long line of Porta-Potties snakes out back. The building itself is a large aluminum shed and inside I discover an impromptu MASH unit. Rows and rows of plywood boxes line the floor and at the end of the rows sit metal tables that the birds are cleaned on. Hundreds of bottles of Dawn, shiny white soldiers, line the shelves in neat rows. Everything I encounter—trash cans, barrels of fish to feed

the birds, towels, and the boxes that hold the birds themselves—is labeled with masking tape and marked as either
"oiled" or "not oiled."

The Cousteau crew now huddles over in one corner with
their bird, filming the intake and cleaning process. No one
seems to be watching, so I wander off on my own between
the rows of plywood boxes. On the first box is a sign that
says "Escape artist—be careful," though I can't see inside to
determine who the avian Houdini is. But it is the second box
that stops me in my tracks. Inside are six pelicans, huddled
together, obviously stunned with fear, their great swordlike
bills pulled into their chests. They came in just this afternoon, according to the tag on the box, and they clearly don't
know where they are, though they know this new place is
terrifying. Their excrement mixes with oil stains on the
white sheet below them and a small tub of fish in the corner
of the box goes untouched. They look too black for pelicans,
and when one stretches out its three-foot-long wing, it looks
more like the dark wing of an osprey or eagle. I stare into
the bird's lightless eyes. It stares back. I have always seen
pelicans as a kind of embodiment of imperturbability, since
they seem so much more stolid than the other birds I spend
time watching. But this bird is clearly perturbed. It makes a
point to keep contact, by wing tip, with another of the enormous birds, its fellow prisoner. It needs to touch something
or someone it knows. Its expression seems to say "What the
hell has happened to me?" It looks scared and confused and,
despite its attempt to cluster near its mates, very alone.

The sight cracks me open.

But I can't stay cracked. I retreat to my brain. I remind
myself that I knew pelicans before they were famous. That
when I first moved to the South, I fell hard for the birds. I
loved their nobility, the way they glided through troughs

of waves, the way they hovered like pterodactyls above me when I walked the beach. As I got to know them better, through books and observation, I learned how much water their enormous gular pouches could hold (21 pints or 17.5 pounds), what they sound like (nothing, they are more or less mute), and even got to see a newborn emerge from its shell (disgusting and beautiful at the same time). Just this week I learned that the Cajun name for the bird is *grand gosier,* or big gullet, a name that proved itself out when Ryan withdrew the huge catfish like a sword. I have had moments when pelicans lifted me out of my own life, usually brief and ineffable moments of delight. This, what I am experiencing right now, is a different sort of moment.

I stay with the birds for a while. My list of pelican facts and encounters means nothing. I look into their black eyes. I hadn't expected this. Like most people around here I have a general sense of the numbers of oiled creatures. Birds are still coming in with some regularity, despite the optimistic spin of the nightly news. At this point about two thousand oiled birds have been collected in the Gulf and about half of those released. But six thousand birds have also been found dead.[32] It isn't these theoretical birds that concern me right now, however, but the ones inside this box. My thoughts come apart again and I feel a sudden and deep empathy with the pelicans. Not birds in general but these six. Jesus Christ ... what would it feel like to be covered in this foreign substance and then to be so dramatically displaced? They shuffle and stare. They try to rest, but fidget nervously.

To have been ripped out of the place you knew as home and brought ... here. I am still staring at the pelicans when I hear a minor commotion in the corner where the Cousteauians are. Apparently one vet, tired of writers and photographers and camera people, decides it is time for

the lot of us to leave. I was surprised by how accommo-
dating the vets were when we first arrived, explaining what
they were doing and answering questions while cleaning
off oiled birds with Q-tips. But now they, or at least one of
them, has had enough. The Cousteau crew have been trying to film the triage being performed on their tricolored
heron, and while they are the most polite and least obtru-
sive of crews, they are now being hustled out. It would
have been nice to film the complete journey of the bird
that they rescued, but you can't help but empathize with
the vets. By this point everyone in Buras is pretty sick of
being filmed or written about. It might be fun at times,
energizing, like drinking six cups of coffee and running
around in a house of mirrors. But the vet is right: enough
is enough. It is time, at least temporarily, to expel those of
us who are chronicling the oiled pelicans, and get back to
the real work of tending to actual birds.

<center>~~~~</center>

Back at the headquarters building I stay up late. Jessica
peeks in at one point and sees the madman who is me
hunched over the computer and typing with two fingers
at high speed. I look up, smile, and wave what I hope is a
benign wave. She says good night and I dive back down.
Plinking the keyboard like a lunatic Schroeder.

"When we try to pick out anything by itself, we find it
hitched to everything else in the universe." This John Muir
quote has served as a talisman and motto for my trip. But
while I've always liked it, it now seems somewhat theoretical. What if this time, instead of nodding vaguely at the
interconnective philosophy, I take it literally? I find myself playing a little game. I conjure up the pelican I stared

at, picturing its black scared eyes, and wonder if it and its situation are really, as Muir contended, hitched to *everything in the universe.* Is this just eco lip service? Bullshit mysticism? How, for instance, is that one pelican hitched to the Russian Revolution or to Saturn or to the invention of soap? I send a mass e-mail out to all the professors at the university where I teach. As I hit Send, I remember that the name "university" is an intended pun on universe, since everything under the sun is supposedly studied there and, at its best, the place should mimic the universe, or at least this world, in its breadth of interests. In practice, however, modern universities tend to operate in ways quite opposite from the "hitched" manner of Muir's quote, putting up walls between disciplines and operating in their own specialized cells. Now I will try to knock down a few of those walls. The e-mail I send asks the professors at my school to hitch the oiled pelicans I saw to their subject of study, and, moreover, to their area of specialization within that subject.

It's a parlor game, sure. And maybe I'm going slightly insane. But could it be true? Could pelicans connect to everything in the world? I turn to work on something else but soon enough I get the first answer to my query. A professor of marine biology replies and I pore over his message. He brings up DDT and writes: "That one pelican, unlucky as it was to meet BP's oil, was lucky enough to be a direct descendant of the very few that survived the DDT era." I know this already and assume his note won't go anyplace new for me, but then I read: "DDT and its relatives were chemicals ultimately discovered by the pursuit of nerve gas poisons for use in warfare, an area in which German chemical labs were especially interested. The Germans, of course, pioneered the use of nerve gas during WWI, and started an 'arms race' with the allies to discover and produce more deadly gases for use

in war. Insecticides are basically nerve gases, so after gas warfare was forsworn by law and practice, the utility of nerve
poisons for killing other species spawned a whole industry."

As a rule I don't drink coffee at night, but that rule, like
many others, is being tossed out the window. I scribble down
a map on a piece of poster board, a web of connections.
This guy is on to something. It turns out that the war-driven
German chemical industry also invented chemical fertilizers, the sort that are used on the farms that are dumped into
the Mississippi and that eventually end up being dumped in
the Gulf, creating the dead zone. And why were these fertilizers invented? Because the standard fertilizer, bird guano
shipped from Peru and Chile, was unavailable, since the
British Navy had blockaded the guano shipments during
the war. The reason for the blockade was that guano was
also used to make explosives. The biology professor continues: "The Germans themselves estimated they would have
to stop fighting within six months of the start of the war
without a new source for explosives, so they turned to their
chemists and asked them to find a way around the guano
blockade. They did. German chemists figured out how to
fix atmospheric nitrogen to make nitrates, which led to the
development of the chemical fertilizer business during
peacetime and allowed Germany to keep fighting. Without
their chemical industry's genius, Germany would have been
forced to surrender early in 1915, before poison gas was
used, and more importantly, before they had to deal with
Russia."

Okay, good, good, but what about connecting my bird to
the Russian Revolution? It turns out he has an answer for
that one, too:

"A German surrender early in 1915 would have allowed
a victorious alliance of Britain, France, and Russia to avoid

the bloodletting of the subsequent war years and left them much stronger. Russia, in particular, did poorly as the war progressed, and the stability of the czar's government, never very strong, suffered as a result. The Germans pushed Russia in several ways. One of them was to make sure that when the first Russian Revolution of February 1917 broke out, which resulted in a relatively democratic thrust in Russian government and a continuance of the war as a major aim, there would be more trouble. The Germans provided transportation to the hard-line Russian revolutionary, Vladimir Lenin, from his place of exile in Switzerland to Russia, where he led the much more violent second Russian Revolution in November 1917, that threw out the czar's government altogether and led to the Russian's withdrawal from the war. So, in a nutshell, if guano (including pelican guano) had not been blockaded, then the German chemical industry might not have existed or at least been as strong and creative as it was, and the first World War might have ended before Russia became ripe for the Bolsheviks, before the invention of processes for making chemical fertilizers, and before the chemicals that led to DDT (that killed off pelicans) were invented."

I can't sleep. My mind is whirling now. Connections and connections and connections. Each one speaking to the incomprehensible complexity of the inventions of evolution—pelicans, for instance, or dolphins, or salt marshes, or chemical weapons, or *Homo sapiens*—and the futility of thinking that we, isolated in our moment in time and limited in our understanding, can possibly comprehend this overall complexity. The fact that we somehow convince ourselves that we *can* comprehend how things work, and that we screw with those workings and then think we can fix what has originally been "built" over eons, speaks to our colossal arrogance. Isn't it obvious that humility, not hubris, is the only reaction in the face of this immense complexity? That shack tops castle in the long run?

We build things and think "They are built well. They will do well unless something unexpected happens." But something unexpected always happens. Isn't that how the world works?

Our engineers tell us "This is what it is like logically. . . ." And then later they say, "Well, no one could have expected this." Well, they are the no ones who should have expected it.

The question is "When are we going to stop calling these things 'accidents'?" How about something less succinct but more accurate, like "the logical result of overreaching and hubris"?

I think of the way that specialization, which has evolved as a kind of necessary defense against all this messy complexity, feeds our hubris. If we are the best chemist or computer programmer or scrimshaw carver we can rationalize our specialness within a small realm. Which makes sense. When we find ourselves overwhelmed by the world's complexity, it is only natural to try to carve out a small corner that we can understand and then call that corner the world. But the oiled pelican reveals that there are no small corners, or that each corner connects to every other. Evolution does not care that the news cycle has passed and that the cameras are turning elsewhere. Its deeper story will continue on after the glare is gone.

It is my last night in Buras. Tomorrow I leave and head back to "normal" life. I lie in bed staring at the ceiling and wondering how—and with whose help—I will manage to connect my oiled pelican to Saturn.

The next day I head over to Ryan's lodge to say good-bye. Ryan is still back home in Lafayette, and the Cousteauians are off on some new adventure. No one is around but Lupe. One more free lunch—a BLT—comes my way. Between bites I ask Lupe how she likes working for Ryan. Maybe I am thinking that I will get a little dirt on Mr. Lambert to slightly muddy and complicate the portrait of this conservative eco-hero.

No dirt is forthcoming, at least from Lupe. Instead she is full of praise.

"A *caballaro*," she calls him. A gentleman.

Slightly less gentlemanly myself, I pop open a Tupperware container and wolf down a cookie. We talk some more and then we hug and bid each other *adios*. I leave behind two notes of thanks, one for Ryan and one for the Cousteaus.

As I head up to New Orleans my car is still throbbing, and, despite Nathan's assurances, the breaking-glass noise grows louder. But I push on. I can't linger any longer. It's time to go.

ATLANTIS

As I drive up from Buras to New Orleans, I consciously follow the route, or at least part of the route, that Ryan Lambert drove during Katrina. At the lodge in Buras he told me the story of those crowded hours. He had been out fishing the day before, back when Katrina looked like just another storm. That was before it hit Florida and started speeding across the gulf on a beeline, then turned north as if it were seeking hasty reservations at Ryan's lodge.

"I've had a bull's-eye on my back for a while now," he said of the oil and Katrina. Never was that bull's-eye quite as prominent as the day the storm landed, with his front door as its touchdown point. He had been in a sound sleep when a phone call from his wife woke him with the news that a category five was on its way. So he loaded up his truck with what was most important, his guns and rods, figuring if he lost everything else he could survive with them. He was ready to leave—telephone poles were already toppling—when he got a call from Mr. Bayle, a Vietnamese man who had worked for him for fourteen years, who said, "My truck's broken."

"This is the real thing, my friend," Ryan said to the man. "You gotta get out of here."

"We old," Mr. Bayle said, adding that their plan was to stay and take what came. Mr. Bayle knew about such things, having spent twelve years in the Hanoi Hilton, three of them shackled to a pipe with thirty other men.

So Ryan drove by and scooped up the Bayles, who carried only tiny satchels and a rosary and a picture of Jesus along with them.

"We live now," said Mrs. Bayle.

They would survive though their house would not. When they returned weeks later, it lay splintered and upside down in the middle of the road.

"People would come back with trucks and trailers to look for their stuff afterward," Ryan told me. "And what they'd find would fit in a baggie."

The stop at the Bayles's house was the beginning of a very long day. When Ryan finally made his way back to his home in Lafayette, he got a phone call from his sister.

"Uncle Rich is trapped in the city," she said. "He's got a broken ankle."

All the cell phones were out because the towers had been knocked down, but oddly the landlines were working even as the water started to rise.

"I'm coming to get you," Ryan said when he called his uncle. "I have no idea when I'm going to be there. It might be midnight or it might be two in the morning. But just know I'm coming. Just go upstairs because if you get in that water you're done."

His wife told Ryan he couldn't go rushing into a flooded city where there was already panic and rioting. And Ryan said "What do you mean? This is what I do for a living. I hunt and I guide. Now I got a license to do it." He grabbed four or five guns and hitched up his boat and threw his bicycle in the back of the truck. He drove down through Saint John the Baptist Parish to Saint Charles and somehow talked a policeman into escorting him across the I-310 bridge into the city. There was no light once he crossed the bridge and to stay above the water he rode downtown on top of the levee. Every time he hit some obstacle on the levee he'd ease the truck down the hill to River Road, in the process driving over and under downed power lines. It was pitch black except for the lights of his truck. Finally he cut up Causeway Boulevard and started heading into the Metairie neighborhood. When he got to the water it really wasn't deep enough to launch his boat, but was too much for his truck. So he grabbed his flashlight and a couple of guns and waded in. On the way to his uncle's house, he saw nobody except for a few national guardsmen up on some high ground. When he finally got to the house, he beat on its side until a little flashlight came on and an upstairs window opened.

"*Bogale,* that you?" his uncle yelled down.

The old man's truck was on high ground and he threw down the keys, and Ryan backed the truck up to the window, where his uncle could climb down along a pipe into the truck bed. They drove slowly out through the submerged and deserted streets, kicking up a wake, and ditched the old truck for Ryan's once they reached it. Then it was back up on the levee. They made it home around four in the morning.

⁓⁓⁓⁓

My entrance into the city is less dramatic, though I do drive beside the levee on River Road, stopping a few times to walk up over the hump of grass to see the river. Brown water sloshes against the green of the hill, a murky algal soup of microbial humus and Big Gulp cups.

As I enter the city it occurs to me that New Orleans is a kind of oiled pelican in its own right. I was joking about gannets being envious of pelicans, but I am dead serious when I say that Alabamans and Mississippians resent Louisianans for getting most of the media attention, and therefore most of the money. And if that's true, then within Louisiana itself the rest of the state resents New Orleans for its place smack-dab in the middle of the national spotlight. I can understand this. This city is a showgirl who never wants the lights to fade. With Katrina this made sense; with the spill less so. While I don't want to take anything away from a citizenry that has endured more pain than Job, the spill seems somewhat more theoretical here in New Orleans than it did down in Buras, despite the obvious impacts on tourism and seafood.

I stay at the the Maison Dupuy, the first hotel I see after pulling into the French Quarter. I unpack and walk down to Bourbon Street where a large policeman grabs me and hands me a ticket. I fall for it for a second, thinking he's a real cop, before I look down and see that my ticket is actually an ad for a topless bar. The street is mobbed but I make my way to a bar called French 75 to meet a friend of a friend. It is wonderfully cool inside the bar after days of sweating, and I discover a delightful drink—fruit, herbs, and alcohol—to savor. As I sip it my mind runs away, imagining the romance of the lonely drinker; the writer tired from his daring travels. When I ask the bartender what I'm drinking, he tells me that it's called "pisco," which turns out to actually be the name of the Peruvian alcohol in the mixture, and sounds rugged enough. It isn't until my third drink that I learn its real name, which quickly bursts my macho pretensions. My new drink is called a "Daisy."

Kristian Sonnier joins me around drink two. An outgoing and generous man, Kristian is a regular here and is therefore a pal of French 75's renowned bartender, Chris, a bald man with thick black-framed glasses who struts about the place in a white suit coat and black bow tie. Chris seems a little full of himself, but I like him because he feeds me my Daisies and then a delicious Cornish game hen and perfect little fries that look like their middles have been inflated with a tiny bicycle pump. I probably weigh about fifty pounds more than Kristian, but I notice that when we shift to beer I start slowing down and he begins to pick up the drinking pace. I have never been to this city before, other than to drive through it on my way to Buras, but maybe in New Orleans I've found my lost tribe of eaters and drinkers. We take our legal "walking beers" through the French Quarter and head down to the river. As we stroll I tell Kristian my theories about the "oiled pelican," the gap between the over-hyped and the real, and he has a funny idea. Before I know it we are off on a celebrity hunt and, soon enough, we find the celebrity in his natural habitat, under the spotlight by the water. There he is espousing about the spill, which the locals find comical since the oil is nowhere near their city. But the locals also love this man and the attention he shines on their metropolis, and that love is apparent as we close in on the CNN truck. Near the truck a small crowd has gathered to watch the white-haired man in a black T-shirt two sizes too small as he delivers his newscast.

Another contact of mine in New Orleans calls Anderson Cooper "the biggest shit stain on the water." I can see how he might fail the "no more bullshit" test, can see the whole phony baloney, superstar, simplistic-take-on-complicated-issues thing.

But Kristian is more philosophical: "Of course it's kind of funny that he's broadcasting from the river, two hundred miles from the real action. But he gives voice to the people's anger. He has Billy Nungesser on quite a lot, for instance."

When he finishes broadcasting, Cooper comes over to where our small crowd stands and shakes hands with the men and hugs the ladies. If there is an edge of Beatlemania to the whole thing, I give the man credit for doing his best to conduct himself with dignity, signing autographs and getting his picture taken and smiling. The only truly embarrassing moment is when some college kids start to slather over him. One particularly enthusiastic (drunken) boy goes on about how much he loves "Anderson," as he calls him over and over to his face, and how he wants to be him when/if he grows up. After he gets his picture taken with his hero, the boy skips off down toward the river, lifted on the wings of celebrity.

That's when I see my own chance.

"I can't profess my love for you," I say. "But how about a picture?"

At which point, just like the college boy, I throw my arm around Anderson.

And so my first night in the city of New Orleans ends with my embracing the king of the oiled pelicans.

The next morning I head out on a bird-watching expedition to the lower Ninth Ward. The neighborhood is famous, or infamous, as the place where the walls didn't hold back the water during Katrina, and it came rushing through in a great wave, wiping out homes. Five years later many buildings still haven't been rebuilt—I drive past plywood houses without roofs and doors—and there is an overgrown, green jungle feel to the roads closest to the water, half-empty lots lush with tall grass and ferns. A black snake with silvery markings crosses the road and marsh grass grows so high it's hard to see around the corner. Up on the wall, which holds back the Industrial Canal, the one that broke and let the water pour through, I spot two yellow-crowned night herons. They are beautiful—black eye bands vivid against white cheeks—and do not seem particularly out of place in the steamy, lush environment.

I, on the other hand, must look kind of funny walking through the neighborhood, a white guy in shorts with binoculars around his neck. I stop and talk to one man who tells me that the water was as high as the tops of the telephone poles. He has rebuilt his house himself, and takes obvious pride in it, escorting me up to his front door. His home stands out next to the tattered and warp-wooded shacks of his neighbors.

On the next street I stop to talk to two women who appear to be moving into a new house. One of the women, Glennis, tells me that Katrina completely destroyed her original home. She is finally moving back from Texas, where she has spent close to five years in a trailer. She points at her new home, which is green and looks a little like an amphibious boat, and, it turns out, was paid for by Brad Pitt as part of his project to restore the lower Ninth.

"I don't like the way it looks," she says. "But I like having a roof over my head."

She also points at the wall that holds back the Industrial Canal and speculates on why it broke open during Katrina.

"They say a boat hit it and cracked it open by mistake," she says. "But I think they blew it up on purpose to flood us out. Like rats."

I nod and let that one sit. When I ask about the oil she does not seem overly concerned.

"I got other worries," she says.

To tell the story of the oil you have to tell the story of the sea. If we are to believe the vast majority of our scientists then today's oily disaster and the rising, roiling seas are all of a piece. They both stem from an obvious enough root cause, our deep dependence on and overuse of oil.

I sympathize with the skeptics. Like Glennis, we've got other worries. Also, could what the scientists say really be true? The world is so large that for centuries we overran it and, try as we might, could do no global harm. Now it turns out that the way we act can actually change the world. Pause on that startling fact for a moment. In our numbers, in our ingenuity, we have found a way to do things that are world changing, in a way we would not have thought possible just a few short decades ago. To most of our minds this notion of large-scale cause and effect seems unfathomable. But, like it or not, we have conducted a science experiment just as surely as I did with beakers and Bunsen burners back in Mr. Grocci's class at Forest Grove Junior High School in Worcester.

I am not in the business of predicting the future, but I will say this: it's only a matter of time until what happened to New Orleans during Katrina happens to another city. Miami is an obvious candidate, but there is another, even more high-profile city that is ripe for flooding. At the end of our tour of the Outer Banks, Orrin Pilkey said something that stuck in my head:

"Let's say the seas really rise seven feet. That's not a prediction, mind you, but a working figure I've now arrived at. If I were in charge of things that is the figure I would use. It's smart to be ready for the worst. The official prediction now is a meter but I think it's too conservative. I would act as if the seas would rise seven feet by 2100.

"If sea level rise really does get to six or seven feet we aren't going to be worrying about a few beach houses," he continued. "We are going to be worrying about Manhattan and Boston."

When I got home from our trip I wanted some con-
firmation or refutation of that seven-foot figure, which
was the highest I had heard. I decided to go directly to
the top and sent an e-mail to Jim Hansen, the head of
NASA's Goddard Institute for Space Studies. Two hours
later Hansen wrote back:

> That's a good figure, in my opinion. If we stick to
> business-as-usual it will likely result in a sea level rise
> of about two meters, which is about seven feet. The
> catch is that if we hit two meters in the order of a cen-
> tury, it means we would be on the way—sea level rise
> would not stabilize at two meters.

Of course Hansen's goal is to dramatize global warming,
and there are a lot of other numbers floating around. The
latest UN report by the Intergovernmental Panel on Climate
Change (IPCC), signed by 2,500 scientists in 2007, predicted
a potential rise of seven to twenty-three inches, but this
prediction is dated by the fact that the IPCC did not consider the melting of ice caps in Antarctica and Greenland.[33]
Since then a chunk of Antarctic ice seven times the size of
Manhattan has calved off from the Wilkins Ice Shelf, and
the predictions have grown more dire.

A few months after my e-mail to Hansen, I talked Orrin
into taking another trip with me. This time we traveled to
an island of a different sort, not a barrier island but a chunk
of glaciated bedrock, home, not to a few thousand people,
but to eight million.

New York City had never felt more primal than it did that
day. Orrin noted how the gridded streets would act as sluice-
ways leading water from the rivers into and through the
city. We walked down into the subway and found a humid

subterranean world. The subway tracks gleamed black, and the white wall tiles dripped sweat as the place radiated a steamy rain forest heat. I tried to imagine even more primal conditions: the tunnels filling with moving waters.

Most primal of all was Ground Zero, a name that takes on a whole different meaning when you realize how close it is to sea level. All those years after the attack and the scene still seemed chaotic. A car ramp led down into a chasm of gray cement walls and Porta-Potties and erector-set bridges and temporary worker trailers and staging and tattered American flags and piles of garbage. This was just about the lowest elevation in the whole city, land that had once been in the water and might be again. I imagined describing the particular configuration of land and water to a geographer, while stripping it of its specific famous and overpopulated locale. What, I would ask, would you call a great chasm less than five feet above sea level that is also less than a quarter mile from a rising body of water? Well, the geographer would answer, I know what I will soon call it: a lake.

In fact, geographers and scientists already have a name for this lower tip of Manhattan, a name that graphically suggests how it might fare in the face of sea level rise. The Basin, it is officially called. One of those scientists, Klaus Jacob, a Columbia geophysicist who is working on a Climate Change report for the city, has gone even further. He calls it the Bathtub. Some New Yorkers used this same name for the excavated Ground Zero site itself, due to its tendency to fill up with water after rainstorms, but Jacob believes the name fits the whole of lower Manhattan.

My brain, like most of ours, tends to focus on the short term. The coffee and doughnuts that Orrin and I had for breakfast, for instance. But as we toured New York I tried to stretch my mind. *Seven feet.* It's a bold, perhaps too-large number. But let's say we accept the fact that Orrin Pilkey and Jim Hansen, who spend their lives studying this sort of thing, know more than we do, that their observations and calculations trump our gut feelings that it couldn't happen, not here, not *really*. Let's just assume, for the sake of argument, that when they say seven feet we should take them seriously. What would that actually mean?

What seven feet would mean in New York is that the Bathtub of lower Manhattan would fill to the brim. Streets would flood, water rushing down into subways that would turn into underground rivers. Standing there that day I felt an odd conflation of disasters. 9/11 melded with Katrina. Strange how our modes of apocalypse shift, like styles of clothes: terrorism, nuclear war, economic collapse. And yet Katrina signaled a shift in which nature itself began to play the role of the heavy. Nature, and of course *us,* the great manipulators of nature. What is sea level rise if not the result of our use of oil and other fossil fuels? It is all connected, both natural and un-. I think of a trip I made to Belize, to a village called Monkey River Town, seven months after the towers fell. Everywhere people talked about the great tragedy that had struck the fall before, but they weren't talking about September 11. The date they kept mentioning was October 9, the day that Hurricane Iris, a category-four storm, had slammed into the coast of southern Belize with winds in excess of 140 miles per hour, killing dozens and leaving 10,000 homeless.

Manhattan is safer than Belize, and safer than New Orleans, thanks to the cooler waters off its coasts, giving hurricanes less energy to feed off. But New York has seen its share of storms. In 1821 a category-four hurricane hit New York City directly, raising a storm surge of thirteen feet in an hour, cutting the island in half, and flooding the

entire city. In 1938 the famous storm known as the Long
Island Express hit the coast with a storm surge twenty-five
to thirty-five feet high. Perhaps most relevant to today is
Hurricane Donna, which struck New York on September 12,
1960 with ninety-mile-an-hour winds and five inches of
rain.[34] The images of Donna help one imagine the storms to
come: people in lower Manhattan trudging through waist-
deep water, others floating along in rowboats. The United
States Landfalling Hurricane Probability Project predicts
that there is an 89.9 percent probability that the New York/
Long Island area will be hit with a category-three hurricane
over the next fifty years.[35] But the truth is that as sea levels
rise it won't even take a hurricane to flood lower Manhattan.
A strong enough nor'easter will do the trick. That is why
hurricane experts see New York, despite the relatively low
odds of a category-four storm, as the country's second most
dangerous major city, behind only the hurricane bull's-eye
of Miami and just ahead of New Orleans. Consider that all
three major New York airports, as well as the rail, and most
obviously the subway, are less than ten feet above sea level,
and storm surge predictions for a category-three hurricane
top twenty feet in most locations. That puts JFK ten feet
underwater.[36]

There are all sorts of plans to prevent this, but they
sound a lot like the usual plans to prevent weather and seas
and winds. Boys with toys again. One plan involves building three large barriers at the Verrazano Narrows, Arthur
Kill, and Throgs Neck, barriers that will theoretically shield
Manhattan in the manner of the Eastern Scheldt barrier
that protects the Netherlands. But beyond staggering costs
is the question of their potential effectiveness.[37] One man
who questions how much good barriers would do is Klaus
Jacob. Jacob, playing the Orrin Pilkey role in New York,

is deeply skeptical about dikes and barriers; he thinks that barriers or walls will just give people a false sense of security. "The higher the defense, the deeper the floods," he has written.

In fact, Jacob has already suggested the same notion to the residents of New Orleans. Not long after Katrina, he caused a stir by writing one of the first papers that proposed that it was foolish to rebuild New Orleans. The idea might have been politically controversial, but Jacob argued that it was also innately commonsensical given sea level rise and the fact that parts of New Orleans are actually ten feet *under* sea level. Why spend a hundred billion dollars to rebuild when the odds are it's going to happen again fairly soon? He wrote: "Some of New Orleans could be transformed into a 'floating city' using platforms not unlike the oil platforms off-shore, or, over the short term, a city of boathouses, to allow floods to fill in the 'bowl' with fresh sediment." New Orleans, he went on, would soon become an "American Venice."[38]

Which sounds nice in theory. But what if you happen to live here?

After my visit to the Ninth Ward, I drive out to Chalmette, a town that was completely submerged by Katrina. Locals talk of how the massive water tank blew off its great stem and bobbed around on the rising water like a giant beach ball. I stop to get gas and start talking to the guy who is sitting on the bench outside the service station. His name is Joe and his house was completely destroyed by Katrina. It wasn't the first time. The same home had been destroyed by a hurricane in 1965.

"That was Hurricane Betsy and my little daughter was

one month old. I remember it was '65 because the next time my house got wiped out my little girl was forty."

He shakes his head slowly.

"If it happens again I'm leaving and not coming back," he says.

We shake hands good-bye. But then as I start to walk back to my car he adds one more thing.

"Of course that's what I said the last time."

On the way back from Chalmette, I see a homemade sign that reads "Car repairs."

I pull into a garage to ask about my shaky car. It's a section of town called Araby, a name I begin to understand as I sit in the garage listening to Arabic music on a boom box while Ali, a smiley, chain-smoking mechanic, fiddles under my hood. His shop is not officially open for business yet, but he offers to fix the car for cheap, an idea that's quite attractive to me. I sit there for over an hour while he unscrews one of the heads and replaces it with a new one, his cigarette resting on the edge of my engine. After a while he shuts the hood and slaps it. "You are good to go," he says. After I pay him—"Cash only"—and pull out of his garage, the warning light no longer flashes and the car no longer bucks.

In the evening I tour the city with a local filmmaker and folklorist named Kevin McCaffrey. We pass gutted shopping centers that look like they could have been used as backdrops in *Mad Max*. I know why I'm thinking of the movie, of course. Again I feel like I'm living in the future. Or is it the past? Five years after Katrina and in some of the malls not a single shop has reopened.

"It's not just the oil that we fear," he says. "The basic math of the Louisiana coast is working against us. The ground is sinking and the water rising."

Food is Kevin's specialty. He talks about the oyster beds

down in the bayou, how they have become less productive. His worry is that it isn't just the oysters that are being lost, but local knowledge.

"The young men file for the same plot that was their grandaddy's. But the coast has changed so much in just the last few years. The old plot may not be a productive plot. Why not put new beds in a new place as the coast moves? Or have they already lost the knowledge to know where to put the beds?"

The oil is a disaster for the oysters, but also for the culture here, since so much of that culture centers on food—"Food is New Orleans," Kevin says—and so much of that food is seafood. In earlier years the Gulf spilled over with a great abundance and out of that abundance grew the food traditions of New Orleans, which hold the city together. The equation is simple: food is New Orleans and oil kills food.

For ourselves, we pass up fancier restaurants and head to a place called Frankie and Johnny's. We eat a good greasy dinner (I have a fine chicken fried steak) and, after we have fully discussed the coming oily apocalypse, the talk turns to football. I'm getting used to this conversational mood shift, characteristic of my encounters here. Burned out on doom, we run to sports. Once again I'm impressed by how much the Saints mean to this beautiful, doomed city.

"Thank God the Saints won," Kevin says. "But that isn't going to save this place."

⁓⁓⁓

A storm is coming, but the people at French 75 seem no more worried than Glennis was about the oil. After a day of exploring the city, I wander back to my new favorite bar for a nightcap. I sip a Daisy on my last night in New Orleans as the customers scoff at the notion that a puny tropical storm like Bonnie could scare people as tough and storm-scarred as they are. There is nervousness under their bravado, however, since you never really know with storms out in the Gulf. Especially when you have just come off the warmest six months on record.[39]

During my recent visit with Kerry Emanuel, the MIT professor who described the "natural human ecology" of the coast, he educated me on these trends. Emanuel is one of the country's leading authorities on the recent intensification of storms. If you were to read a so-called balanced account of this issue in the newspapers you might come away believing that the scientific jury is still out on whether or not our warmer waters lead to more intense storms. In fact, this is a little like saying the jury is still out on evolution versus intelligent design. The real split, Emanuel explained to me, was not in the scientific community, but between the scientists and the weather forecasters. He assured me that what common sense suggests is true: warmer waters lead to more violent storms.

People talk tough in French 75, but they all know that this will not be the season's last storm. And as for the Big One, it's only a matter of time.

"All that oil is out there somewhere," says a man down the bar. "A storm would churn it up and send it raining down on us."

I have no idea how feasible this is. This was one of the early legends of the spill. You heard it a lot down here, that there would be oil raining down on New Orleans. It sounds more like a biblical plague than science, and maybe that's the point. Is fire from the sky next? A host of beetles? Nothing would surprise the citizenry of this town.

My mind has migrated to a strange place. I'm falling for

this city, and I'm having a great time. Bring on the Daisies and the Cajun food. But at the same time a part of my brain is filling with thoughts of disaster. It's not just the oil or Katrina. These simply seem a part of a larger series of disasters to come. Things fall apart, and they have, and are, at an alarming rate. I don't really know how to respond to this since I am not, by nature, an apocalyptic thinker. But maybe I happen to be living, by pure bad luck, in apocalyptic times.

At the end of my long hike through New York City with Orrin Pilkey, he climbed in a cab and headed back to the airport while I walked south for a couple of miles to do something I hadn't done since I was a kid. When I got to the Empire State Building I waited in line for well over an hour—first the line for tickets, then the line for the first elevator, then the line for the final ride to the top. But the wait was worth it. The weather was clear and the view was both startling and scary. How could human beings build anything so high? I snuck out within ten feet of the edge and looked toward downtown, toward the Basin, the Bathtub.

The city was laid out below me like a map. Back on the Outer Banks, Orrin pointed out the folly of having streets that ran from the ocean to the marsh side of the barrier islands, since during storms those streets could easily turn into inlets. New York's streets would perform the same function, leading water from the Hudson to the East River, *ushering* the storm surge straight into and through the city, crisscrossing the place with water and helping flood the island.

It was hard for me to get my head around this picture of a drowning city. How can we not be a little skeptical? *This is the way the world is,* we think, *the world we know, and this is the way the world will stay.* To say that most of us are climate change skeptics is not to say that most of us doubt the work of science. What we are perhaps truly doubtful of is the ability to predict or believe in radical change. It's simply not in our DNA.

From the Empire State Building I looked down and, tired of picturing what might be, I imagined what once was. Were I able to travel back in time, say a thousand years, the island below would have been crisscrossed with streams and spotted with marshes, not cross-sectioned by streets and studded with buildings. The site where the Trade Center towers once stood would have been well offshore to the east, submerged then as it might be again. The only human beings that inhabited the place would have most likely done so seasonally, the way most humans have always used the coast, taking advantage of summer's abundance but wary of winter's cold winds. The island itself, like all islands, once migrated, forming and reforming through storms and currents, though that was before buildings were shot down into it, like framing nails, to try to keep it still. But of course it won't keep still. The human desire to pin things down won't keep the place from changing: the island was once something else and soon enough will be again.[40]

Up there, on top of the world, I found myself growing weary of apocalypse. Despite Orrin Pilkey's energy and charm, I was still not yet entirely sold on the fact I so often hear repeated: that global warming and sea level rise will lead to our doom.

But then again, it didn't hurt to play a little game of "what if."

What if? It was easier to picture it from above: water filling the lowest areas first, pouring into the Trade Center site, cascading down the subway steps, using the cross streets to cut inland toward the island's middle. Even with a mere .71-meter sea level rise by midcentury, the conservative number the city planners have settled on, New York faces obvious threats to its water supply, sewer, and wastewater systems, coastal erosion, and saltwater infiltration of aquifers and surface waters. Meanwhile, according to the United States Global Change Research Program's assessment, "subway, road or rail tunnels or ventilation shafts will be at or below flood levels."[41]

And this was without a hurricane. There's the rub. The fact that gets lost in the squabbling over numbers is that sea level doesn't have to rise a single millimeter for a Katrina-like disaster to strike New York.

While I had spent the better part of the day trying to imagine this, the truth was that the word *unimaginable* worked pretty well in this instance. Eight million people on an island that is close to sea level with questionable means for evacuation. Experts can talk in a positive fashion about "solutions," but the truth is that while sewage barriers are nice, the current strategy really comes down to being lucky. After all, isn't it simply due to good fortune—a kind roll of the dice—that storms have veered in other directions and that a major storm hasn't hit in at least a decade? We can create barriers for sewage plants—and by all means we should—but we can't control what is uncontrollable. No one wants to hear it but the real conclusion of an honest climate change impact report would be pretty simple: keep your fingers crossed.

I pictured a storm barreling up the East Coast and then veering northeast, drawn into the funnel of the Verrazano Narrows through the Upper Bay and toward the Hudson and East rivers. Once there, it would send twenty-five-foot storm surges over Battery Park and wash over Ground Zero, filling the bathtub of lower Manhattan to the brim.

I am no prophet, and I certainly wasn't saying this *would*

happen. But there is no doubt that it *could*. Or it could do any number of other things, could veer slightly east and take out the Rockaways, for instance, or slam into Coney Island, once a healthy barrier island before its leeward marsh was artificially filled, where high-rise buildings now stand right next to the Atlantic. Or the storm could decide—or whatever the storm equivalent of "decide" is—to skip New York entirely and head for Boston or Washington.

Like most of us, I could barely imagine this happening. But that, I understood, had no impact on whether it would or wouldn't.

For now New Orleans retains the title of the country's most primal recent disaster. As I walk back to my hotel from the bar through the still active streets of the French Quarter, there are few signs that this is Main Street, Atlantis. We are resilient animals—of that there is no doubt. We shake off our unimaginable disasters and move on.

I decide to take a detour down to the river. Tomorrow I am supposed to head home and if good sense were my only guide I surely would. But a part of me resists. There is one Gulf state I haven't visited yet and I am feeling an irrational urge to head, not back to North Carolina, but to Texas. Specifically, I want to see the city of Galveston, a city that drowned a century before New Orleans did.

IV. Beyond the Oiled Pelican

MIGRATIONS

Fall is coming and the world jitters with movement. Right now up in Minnesota loons are stirring. Soon they will leave the Northern lakes for which their calls serve as a signature and fly down to the Gulf for the winter. These sleek, dark diving birds will arrive by the millions and will plunge deep into the oily Gulf. Meanwhile Ryan's ducks are already on the move, blue-winged teal and scaups among them, and by late August they, too, will pour into the Gulf.

The early waves of birds are sweeping through, an advance wave of a massive exodus that will funnel down the great corridor of the Mississippi before making the bold move across the Gulf, linking the northern hemisphere to the southern. "Bird migration is the world's only true unifying natural phenomenon," writes Scott Weidensaul, "stitching the continents together in a way that even the great weather systems fail to do."[42] The birds might as well trail thread behind them and hold needles in their bills, so obvious is this stitching.

Before I left for this trip I talked to as many bird people as I could about the coming migration.

"Imagine running a marathon in bad air," said Laura Erickson, of the Cornell Lab of Ornithology. "Well, for many species this is their marathon and conditions need to be just right."

For some birds the marathon includes a six-hundred-

mile water crossing over the Gulf. Which means a marathon following the supermarathon that is the overland
central corridor migration itself. Over a billion birds flying down the middle of the country, a migration contain-
ing not just a stunning number of birds but also a stunning
variety.

Migration is always a gambit where everything has to be
perfect, birds building up body fat from high-quality sources
of food at certain stopover points. Imagine if the fish a migrating bird eats is compromised, or if a certain sandpiper
can't find a certain crustacean, or an abundance of certain
crustaceans, where it has reliably found them before.

Of course none of the scientists I've talked to claim to
know exactly what the results of the spill will be, either
short-term or long. They are cautious, and everyone I speak
to is understandably quick to use caveats and qualifiers,
which leads to other questions. Not only do we not know
the results, but we may not know them for years. How will
we gauge the ultimate effects of the spill?

Back in Alabama, at the Dauphin Island Sea Lab, I spoke
with Senior Marine Scientist John Dindo. His was a brand-
new building but the air-conditioning had died on July 1 so
fans blasted us as we talked.

"There is a whole lot of unaccounted oil out there and
what happens to that oil is the big question."

He adjusted one of the fans on his desk so that it pointed
straight at me. I was grateful.

"I'm not too worried about animals. Animals are highly
resilient. For the most part they can repopulate an area in
a short time. But habitats are not resilient. Habitats that get
affected by the oil cause a cascade effect. All the way up and
down the food chain."

But Dindo stopped short of pronouncing that the oily

apocalypse was upon us. He was reluctant to make any grand conclusions and was skeptical of those who were doing so.

"Scientists should not be making bold, definitive statements at this time. It is tempting, but what we should really be concerned with is not today or tomorrow, but a potential decadal situation. Anyone who makes a bold statement at this point doesn't know what they're saying."

I liked this. BP's great flaw was certainty, and scientists and environmentalists should be wary of mirroring this flaw. It seems to me that the only real conclusion at this point, the only truly scientific conclusion, can be summed up in five words: we just don't fucking know.

<hr />

This morning I begin a migration of my own. I am not driving home, however, not yet. What I *am* doing makes no sense really. Yesterday my wife and I were supposed to close on our new house, our first after years of migrating from north to south. We've decided to settle on, and in, a home off a salt marsh in North Carolina.

The seller is upset that I'm not back and is threatening to pull out. I swear to him and the Realtor and to my wife that I'll be back, even if I have to hitchhike or crawl. I *will* be back Monday, I tell them, though to be honest I still hate the thought of leaving this place. Maybe it's just that end-of-a-trip feeling, but I decide that before I leave I need to make one more stop. And so, rather than immediately heading north and east as I should, I point the car west toward Texas. My goal is to get to Galveston. The trouble is that it just happens to be eight hundred miles in the wrong direction. I drive fast and turn up the music.

One reason I'm heading to Galveston is that it almost

perfectly embodies the theories of the straight-line thinkers like Orrin Pilkey's bête noire, the Army Corps of Engineers. The story of the city, its early drowning and recent re-drowning during Hurricane Ike, is as simple and powerful as a fairy tale, though a grim old-fashioned fairy tale of the sort New Orleanians might read to scare their children. Or you could imagine Orrin Pilkey taking the owner of some coastal condo on his knee and telling the story in a fright-ening whisper. Either way it's the story of a people who be-lieved in straight lines, a story of a people who built high walls, thinking they could stop the sea, and a story of how those walls ultimately doomed the city. In the summer of 1900, Galveston, which was then a thriving port town, was struck by a storm that remains the deadliest hurricane in United States history, killing 20 percent of the city's people and destroying all of its buildings. The town responded by vowing "never again" and constructing a massive seawall. But the wall has ultimately acted, as geologists like Orrin and others have warned it would, by destroying Galveston's beaches, its natural buffer and protection, leaving it beach-less and therefore defenseless. Which led to this decade's di-saster when Ike struck and wiped out the city again.[43]

By heading this way I'm also following the refugee route of those who left New Orleans to spend their years in the diaspora of the Lone Star State. Cajuns among cowboys. Like Glennis of the Ninth Ward, they passed the years in trailers while back home their houses lay underwater or in soggy tatters.

Homes and homelessness are much on my mind. Back when I lived on Cape Cod, John Hay talked to me a lot about the idea of home. He had lived in the same place for almost sixty years and he spoke of our need to "marry" the places where we live, to spend a lifetime learning the land

and people. He also believed that many of our national troubles spring from a "great epidemic of homelessness."

"People are on the run everywhere these days," he told me. "As if they don't know where they live. Everyone seems intent on dispossessing themselves."

As I travel, I can't help but feel this is true. Most of us don't know where we live and those who do know are uprooted by circumstances beyond their control. We move and float, rootless and out of place. We have become a homeless nation, drifting, uprooted, unsure of where we are.

I want to be wary, however, of mere theoretical homelessness. At a time when more people are losing their actual homes, I need to watch my metaphors. I think of my brother, who was homeless for a while on the streets of Austin. His time on the streets affected him in many ways, but one of the smaller consequences was literary. He could no longer stand to read Jack Kerouac.

"I read *On the Road* when I was in a homeless shelter in Texas," he said. "I hated it. Kerouac was *playing* at being a bum. He doesn't know anything about it." Rootlessness was no longer a playful thing to him.

※

In the book *Turtle Island,* Gary Snyder exhorts his readers to "find your place on the planet. Dig in, and take responsibility from there."[44] Two people who have taken his advice—to dig in and fight, in Mobile, Alabama—are Bill Finch and Bethany Kraft. They both impressed me deeply on the day we saw the great sand mounds of Dauphin.

Earlier on that same day, Bill had taught me some things about migration by reminding me that the invisible is also part of the so-called real world. We drove into Bayou La Batre and turned up University Road, which within seconds happily devolved from a paved street into a bright orange-red clay road. We stopped to examine a small body of water on the roadside, more puddle than pond. Bill brought up the subject of carnivorous plants, and I boasted, with newfound regional pride, that my current home in southeastern North Carolina was a kind of unofficial capital for bug-eating plants.

"Yes," Bill admitted, "you have almost a dozen species of carnivorous plants."

And then he dropped his naturalist's hammer. "Here we have over a hundred. *This* is the world's capital."

He went on to prove his point as we hiked into the longleaf pine savanna, pointing and giving Latin names until I was ready to cry *"No mas!"* The plants were beautiful, especially the tiny sundrop, bringing dewy death to bugs who thought they had struck water, as well as a larger red-veined beauty that looked like it took pleasure in its work. Bill kneeled down by a pitcher plant and broke it open to show me its innards where an insect was being digested. "Darwin said he cared more for his work with carnivorous plants than all the rest of his work combined." You could see why: this was a miraculous and miniature world that almost no one was aware of. Before we left the savanna, Bill offered each of us a piece of toothpick grass, a spirally, brownish-beige grass that numbed our gums.

But the real treat was just ahead: A couple miles down the road we climbed out of the car and cut through a slash of pine forest and into the marsh itself. We walked through grasses that were up to our waists, with no way of seeing where we were planting our feet. The only thing keeping us from stomping on a cottonmouth was luck, that and the fact that maybe we were loud enough for them to slither off and avoid us. Of course I didn't utter a peep about my fears.

Birders and other nature types are regarded as somewhat geeky by the general populace, if they are regarded at all, but we have our own sort of unspoken machismo rules.

I can say with confidence that I've spent more time tramping around marshes than most people, but I usually stick to the mucky earth. Not Bill. When he reached the marsh's edge he waded right in without hesitation. He, like Bethany and me, was wearing long pants and sneakers, but soon he was up to his knees, and then thighs, sloshing through the water of the salt pan that bordered the taller grasses of the marsh. We followed suit, each of us splashing along, fanning out into different parts of the landscape. Bethany stopped to take pictures and I followed Bill for a while before drifting off and staring out at the endless prairie of tall grasses.

We were walking into one of the most miraculous inventions in the history of the planet: the salt marsh. The salt marsh, a great nursery where almost every fish in the nearby sea is born. The salt marsh, whose moods change every six hours or so, transforming it from a rushing river to a landscape of stranding muck. A true edge between land and sea, fresh- and saltwater, forest and ocean.

"Alabama the Beautiful," the license plates say, but I wasn't buying it when I first drove into the state. Alabama was toothless folks listening to Skynrd and getting frisky with their cousins out by the still, and their landscape was as scraggly and dirt poor as they were. But here was the real Alabama: the grasses soughed in the winds, and the Gulf beyond roiled and crashed against the marsh, and the birds—egrets, herons, sandpipers, ospreys—shot overhead, and other than our small group there wasn't a human being for miles and I was starting to feel that old familiar exhilarated feeling.

Just then Bill called out. He'd struck oil. We all sloshed over to him. And there it was. Not the goopy black variety of our nightmares, the kind I'd imagined before coming down here, but a light blue sheen, beautiful really if it weren't for what it was. The oil weaved and curled through the marsh grasses. Bill touched it and rubbed it in his fingers and it left a rusty red film. I did the same.

Bill was cautious. Marsh plants create their own natural oil, a kind of vegetable oil that has a sheen much like what we were seeing. After all, Bill stressed, oil also comes from plants and the marsh is adept at breaking down oil.

"Oil isn't alien," he said.

But the volume and residue argued that the oil we were seeing was from the spill. And it had worked its way into the deepest recesses of the marsh. When Bill and I talked by phone, back when I was still in North Carolina, his voice sounded pained when he considered the prospect of a great volume of oil smothering the marshes, the home for so much fish and bird life, and the stopping point for migrating birds. We worried that the very complexity of the landscape might doom it; that while the oil might roll off a beach, or be cleaned by humans, this would be impossible in a marsh, where the variety of grass—cordgrass and needle-rush and eelgrass and other spartinas—and the terrain of mud and muck are built for trapping nutrients, and would therefore trap oil as well.

Though the oil we had found was troubling enough, we were both relieved that it wasn't thicker. Psychologically, it was an interesting moment for us. That earlier phone call had been full of emotion. It was a different world then, right after the oil started spilling, and we imagined gooping Valdez quantities of oil smothering the marsh. Though we, as environmentalists, have something invested in resisting the notion that "it's better than we hoped," the fact is that it was now better than we hoped. And yet how much of this was due to the dispersants? We didn't know. What was the price we would pay in the future? Again, we didn't know.

Bill cut through the taller grasses to the shore of the Gulf itself. I followed, thinking a machete would have come in handy. When we finally pushed through to the Gulf waters we were relieved to see no obvious signs of oil coming in with the tide.

We splashed back to the salt pan and Bill pointed to tiny fish below the surface—sailfin mollies—with backs that flashed a blue that mirrored the color of the slick above them.

While Bill was glad we were looking at a sheen, not a thick black slick, he was still worried.

"People want to see dead animals," he said.

I didn't understand him at first.

"They want an obvious symbol of the devastation to rally around," he continued. "And that means dead, oiled birds. But what we are likely to get is going to be a lot more subtle than that."

What we are likely to get, he went on, are invisible changes that will make it harder to match cause and effect. He pointed to the brown and white periwinkles that clung to the tops of marsh grasses.

"How many periwinkles do you see?" he asked me.

I looked around.

"Millions," I said.

"That's right, millions. And while they look nice they are really trying their best to swallow the whole marsh. They actually have a weird way of eating plants. They slice them open and introduce a fungus and then they eat what the fungus digests. And the marsh supports these periwinkles in huge numbers. But to support them the marsh has to be growing at full speed and if it slows down the periwinkles devour it. What happens when the oil gets in here, and begins to take the oxygen out of the soil, and suddenly the marsh periwinkles, which the marsh had supported just fine, start swallowing the marsh. Then consider what happens if this year's crab population—the young are somewhere out in the middle of the Gulf right now, just starting to migrate back—gets whacked by the oil. So we see fewer crabs over the next year or two. Guess what is the only significant predator of periwinkles?"

"Crabs?"

"Crabs. It's not as obvious as a fish floating belly-up, but it's one way for the marsh to die. And then two years from now when the marsh is dead everyone says periwinkles are to blame. But, oh no, they're not. Things have been thrown out of balance."

Why did it really matter if the marsh died? After all, there wasn't another human being in sight for twenty miles or so, and most of those humans were clustered over in Bayou La Batre, puttering out to sea in Vessels of Opportunity. Well, one reason it might matter was that the marsh was both birthplace and preschool for almost all the fish that, in better times, those same boats harvested. In other words, whether they knew it or not, the townspeople depended on periwinkles.

Complications multiply. A damaged marsh would no longer support sailfin mollies. Which could mean that a green heron, exhausted in its migratory journey and eager to reach what it knew would be a bountiful stopover point, would go hungry.

I drive through a sunken land, at times crossing more bridges than regular road. I don't know if I am getting across how attractive this Gulf landscape is, how visceral its appeal is to a man who loves birds and water. By even the most conservative of scientific estimates, this place won't exist—or rather will exist underwater—in two hundred years. From the looks of it, it could happen in two. It's so low, so marshlike, so permeable, so inconstant. In other words, perfect. It seems to me an accurate map of what the world is really like, even if you live in South Dakota. These days you don't need water around to know that things will be changing soon, likely tomorrow.

The world is on the move and so we build things with straight lines—highways, toasters, oil derricks—anything to stop it from moving. But wait, here's an idea: what if instead we just accepted that things are constantly in movement?

Easier said than done. Perhaps I am a better preacher than a doer. All morning, as I drive toward Texas, I've experienced a kind of free-floating anxiety. In my cupholder is the blue EPA stress ball that I got at the meeting in Buras, and I take it out and squeeze it a few times in my right hand. Then my vague anxiety becomes less so, specifying itself when the dashboard lights up again. The little illustrated engine catches fire and blinks red. I curse Ali, though not Allah. What the fuck? Now my house and hearth and home, and wife and daughter, feel farther away than ever. My anxiety spikes and won't go away no matter how I squeeze my ball.

And so finally good sense, a rare visitor to my mind over the last few weeks, gets the better of me. It is time to turn around and head home. Galveston was my goal but now the town of Grosse Tete, Louisiana, becomes my end and westernmost point. In Grosse Tete I find a ramshackle little building called the "Doc-Your-Dose Pharmacy" and I chat with an older white woman by the river who assures me that the destruction of the mostly African American lower Ninth Ward was "God's will." Then I find a mechanic whom I somehow trust despite the fact I can barely understand his accent. He assures me that what is causing the light is the catalytic converter, and he thinks I can make it home to North Carolina.

"Yull makeit," he drawls.

So the day, which began with grand intentions, becomes what amounts to a 160-mile U-turn.

On the return trip, I almost manage to show great discipline and drive right past New Orleans. Until something makes me tug the wheel to the right and, before I know it, I'm taking the last exit and bumping back down through the narrow streets of the French Quarter in search of French 75 and one final Daisy. To counteract or complement (depending on your take) the drink, I also request a road cigar, and a bartender named Cindy (who seems every bit as professional and a whole lot nicer than the famous, haughty Chris) promises me that the cigar she clips and hands me will be a "long burn," which it indeed proves to be since I smoke it, on and off, for most of the way home.

⁓⁓⁓

Before we left the marsh on Grand Bay, I took a break from gloom and doom to just watch the birds. Willets—large, elegant sandpipers with sharply patterned underwings—flew by and let go with crazy seesawing cries. A tricolored heron fished patiently. Its colors startled: white throat and caramel neck stretched up high against the backdrop of the vibrant green marsh grass. Suddenly it stepped forward and its blade of a bill flashed and the life of a small fish ended.

Bill pointed down at a green, wormlike plant called glass-wort, which has a celery taste. He described a miniature tomato that actually tasted more like tomato sauce than a tomato itself, as if it had already been salted and cooked.

"We would come here and mix that up with the glass-wort and some wolfberry and then get a couple of oysters and have a great lunch. It was delicious. Savory."

No one needed to point out that we wouldn't be lunching on marsh plants today. We looked out toward the sea, where ospreys and gulls hovered, and where the young fish of the marsh would eventually migrate. Then we hiked back up to dry land, walking out through a forest of charred bark like turtle shells and through long grasses and plants called rose pink that made the place look like the poppy field in Oz.

We gradually made our way back to the cars and headed out the red clay road. The road, Bill explained, served as a dividing line between freshwater and salt.

"I wish they'd take it out," he said. "It acts as a causeway and doesn't allow for interaction between fresh and salt-water habitats. Without it you'd have a lot rougher edges between the two.

"You have to listen to what the marsh wants," Bill contin-ued, pointing at the way the road blocked the proper flow of water. "This marsh has deep connectivity issues."

It occurred to me that those words applied to me and to all of us. That I—and Bill, and Bethany, and the tricolored heron and BP and our country and the sailfin mollies and the periwinkles and the crabs—have deep connectivity is-sues, too.

·····

I am not the only writer to be leaving these parts. The mass migration has begun, reporters taking their pens and cameras with them, off to the next place. You can feel the great national spotlight swinging away, the news cycle ending. We are all tired of oil. Time for the next disaster.

As the writers leave, the birds arrive. I picture the millions of migrating birds streaming through the dark overhead, toward the degraded marshes and waters.

Before I left the Buras headquarters, I decided to call Scott Weidensaul to talk about the fall's migration.

"It's hard to think of a species of migrating bird east of the Rockies that doesn't fly through the Gulf," he said. "And these birds, already stressed, are going to be flying into uncertainty."

He pointed to the ways that migrations were already strained. Even in a non-oiled year, new development could ruin a bird's chances. Imagine returning to a copse of woods where your kind had landed for generations only to find a mall or housing development.

"We have been hearing that the worst of the oil is over," Weidensaul continued. "But that oil went somewhere and those toxins went somewhere and that somewhere is the food chain."

He pointed to an example that was quite similar to the one that Bill Finch described back on the Grand Bay marsh. While Bill talked about crabs and periwinkles, Scott talked about whooping cranes. Whooping cranes feed almost exclusively on blue crabs. Blue crabs, as it turns out, have been one of the few species that have been tested extensively at this early date. What scientists have discovered is that almost 100 percent of the crab larvae have traces of oil and dispersant.

"It will be very hard to tease out," Weidensaul continued. "To connect cause and effect. To say this is clearly because of the oil spill."

I have come to better understand what Bill Finch meant by saying that people "want to see dead animals." The reporters and cameramen have focused their cameras on single birds, covered in oil, unable to fly. It is a powerful image, beamed all over the world, and pointedly tells a story of the tragedy of many embodied in one. But it is also a mistelling. Cameras can't tell the larger, deeper story. That's because nature is not merely a series of connections, but also a series of perfectly timed connections. Like a symphony. A symphony whose invisible conductor is time, time on a level humans can't imagine. When that conductor points its baton, the trumpets blare or the flutes sing.

Phenology is a word coined by naturalists for nature's impeccable sense of timing, for the way that, as the year progresses, a fish will spawn in April just after the smaller fish it dines on have spawned, or the swallows will return north just as the insects appear. One of the climaxes of this great symphony is migration. As bad as it is right now on the shores of the Gulf, the coming migration promises a different level of impact. Birds from the north may land at a certain time to find what is not there. We can't and won't know the results of these journeys, since they are largely invisible to us. But we worry that the music may stop.

There are already tears in a web that took millions of years to create. If the whooping crane can't eat crabs, they either can't fly or do so severely weakened. If the crabs don't prey on the periwinkles, the periwinkles eat the marsh. And while this may not be the sort of disaster we first envisioned, that does not mean the results aren't disastrous.

At around eight o'clock, still puffing on my cigar, I drive into Hattiesburg, Mississippi, for dinner. I push on and sleep in a hotel in Meridian, Mississippi. I drive all the way through the next day, my thoughts dulled by the road. Migration barely enters my mind, except once when I pull up behind a truck with oversized tires and Georgia plates and a bumper sticker that reads "Hey Audubon, identify this bird." Blearily, I obey, trying to make out the creature with the swanlike neck, only gradually realizing that it's a middle finger.

I arrive home on Saturday night. A glorious reunion with my wife and little girl. On Monday it turns out the seller of the house has calmed down and we sign the papers. We now own our first home. The next three weeks are a marathon of moving all of our belongings to start our new life by the salt marsh. For once we will be settling for longer than a single year, though we are also well aware that hurricane season is now in full swing and so have few illusions of permanence. Still, when I am out in my new backyard I feel like I've found a good camping spot, only this is a camping spot where we can stay for years. Fairly quickly we add a second home when I build a Hadley, a fort at the base of a magnolia tree.

One day a friend helps me move my two kayaks from my old place to my new. We do this by water, paddling along the Intracoastal Waterway and camping on a dredge spoil island for the night. I watch the blazing orange ball rise near the same spot where the moon—red also, but darker and not as fiery—came up the night before. The next day I paddle home, and see a bittern, usually solitary, in the reeds behind the house. Kingfishers ratchet along above the water, occasionally stopping in midair before diving. I eye the spot out on the marsh where I will soon build an osprey platform so that the birds can nest here in the spring. By the time we have settled into the house the signs of migration flow through the marsh: the early swirls of tree swallows and Canada geese honking overhead.

Sure enough the first hurricane isn't long in coming. I

have been keeping my eye on the weather, wondering if a storm will stir up the oil in the Gulf. But this storm veers northward instead of west and for a while we are in the bull's-eye of Hurricane Earl, briefly a category five. Earl slows and stays out to sea and though I dutifully bring in the lawn furniture, barely a pinecone stirs in our yard. So far we have been lucky, as has the Gulf. There is talk of dodging a bullet followed by much knocking of wood.

It feels bizarre to settle after the unsettledness of the summer. But in the face of coming storms, in the face of a country caught in a crisis of homelessness, and in the face of my own uncertain self, I begin to root down.

"The first flush of rootedness can't be repeated," a friend on Cape Cod once said to me.

All fall long I am deep in that flush. While I've never owned a house before, I have spent my whole adult life dreaming of having a place to call home. It is strange that that house turns out to be in North Carolina, but less strange that it turns out to be on a salt marsh. Not only is the marsh a miraculous ecosystem where I now daily hear the applause of clapper rails, but it connects me by water to the many other coastal places I have come to love over the years. I don't mean this mystically, but practically. If one day I am feeling particularly ambitious, I can hop in my kayak, hang a right, and paddle south to Florida before hooking around to Ryan's lodge or Anthony's fish camp.

As we begin to settle here, I think often of John Hay and his neighbor, Conrad Aiken, the Pulitzer Prize–winning poet. I dip into Aiken's *Collected Letters* and learn that Aiken and his wife, Mary, moved to Cape Cod in 1940. On May 21, 1940, the day the Aikens bought their house in the town of Brewster, Conrad wrote to Malcolm Lowry:

Ourselves, we pick off the woodticks, and pour another gin and French, and count out the last dollars
as they pass, but are as determined as ever to shape
things well while we can, and with love. Nevertheless,
I still believe, axe in hand, I still believe. And we will
build our house foursquare.

The rest of the letters from Brewster are the sort of combination of pastoral and grumble common to those tackling
renovating an old house in the country. Conrad spent his
time "weeding the vegetable garden, mowing lawns, cutting
down trees, shooting at woodchucks and squirrels, attack-
ing poison ivy with a squirt gun," as well as "scything the tall
grass," and, as usual, drinking copious amounts of alcohol.
The poet, then fifty-one, had a good deal of pride in what he
and Mary were accomplishing—"We both thrive on hard
physical work, and feel extremely well"—and exalted in his
new surroundings, surroundings that would soon make
their way into his best poetry.

To shape things well while we can. I write Aiken's line down
on a note card and tape it above my desk. It seems to me as
good a credo as any in these uncertain times as we try to
make this place our own.[45]

DÉJÀ VU: A NORTHERN INTERLUDE

One of the highlights of the fall here in North Carolina is the northern gannets' return to our coastal waters. Northern gannets are long white birds with wing tips that look as if they were dipped in a pot of ink, and they get their dinners by diving spectacularly into the sea. During the cold months hundreds of the birds hover far above the water before pulling in their wings and descending like great white arrows or lawn darts, straight down and fast, accelerating as they plunge into the water. Then they enter another world, turning into loons, tunneling below after fish. You would think this dramatic expenditure of energy would leave them in need of a rest, but they seem tireless. They dive again and again, never flying to shore to rest like a gull.

For me it was a pleasant surprise to find gannets in North Carolina. During our early months in the South I discovered wild new birds—ibises and black skimmers—but my first sight of gannets in early November, when the heat finally lifted, was different: not a new acquaintance that delighted with novelty but an old friend showing up in unexpected surroundings. I'd always thought of gannets as an exclusively northern bird—it was right there in their name—a bird that would barely deem to drop down to Cape Cod Bay from their rock homes in Canada. I had never taken the necessary minute to skim my field guide and notice that in winter gannets actually migrate as far south as the

Gulf. A cartoon double take was required that first day I saw them out over the ocean. Could it be? It could and it was. Suddenly there they were: the familiar white arrows plunge-diving into the surf. These cold weather birds, these *northern* birds, I had always falsely thought, were hurling themselves into southern waters. It was one of my first moments of feeling at home, of thinking that the South might be a place where I could actually live.

One peculiarity about gannet migration is that, being offshore birds, they never cross land. Which means that they arrive in Carolina waters later than most birds, and that they arrive in the Gulf even later in the year. I love the way gannets dive, wildly, deeply, and with abandon, and I worry for them this coming winter. They will be out there where the oil is, diving down to the spots where the great shrouds have been discovered by scientists. They are, for all their abandon, a somewhat fragile bird, and I once walked through the dunes on Cape Cod and found several dead after a nor'easter.

And I now remember a small fact that looms large: the first oiled bird found was not a pelican, but a northern gannet. In fact last spring they were one of the hardest-hit birds. Seabird expert Bill Montevecchi, of the Memorial University of Newfoundland in Saint John's, points out that while it was once thought that only 5 percent of the Canadian gannet population wintered in the Gulf, the number is actually closer to 35 percent. Thousands of gannets will spend the coming winter down in the Gulf of Mexico and, unlike most birds, they will spend it far offshore and diving deep.

John Dindo, the scientist at the Dauphin Island Sea Lab, fretted when we talked.

"I'm most worried about the shorebirds and diving birds,"

he said. "The gannets and ospreys and terns and loons that are diving down into it."

What they are diving down into this fall is the great unknown.

While I was in the Gulf I received an e-mail from Eva Saulitis, a woman I met three summers ago in Homer, Alaska. When we met we had a drink in the bar of the Land's End hotel while staring out at the beautiful Kachemak Bay. I'd barely unpacked my bags at the hotel, where I was teaching at a conference, when I saw a minke whale rise of out of the water, and over the next twenty-four hours I witnessed sandhill cranes gliding and stretching overhead and was startled by an equally startled moose. As we sat in the bar we could see white mountains jutting up across the bay, a sight that could have come off a calendar or out of fantasy. But the view wasn't entirely pristine. Down on the beach a bald eagle, our nation's scavenging symbol, ripped apart a halibut tossed to it by tourist fishermen, who snapped its picture, blood on its beak.

Eva had spent the last twenty years studying killer whales out on those same Alaskan waters. She had become intimate, or about as intimate as a human being can be, with a particular group of whales, a small population of genetically unique, mammal-eating killer whales that live in Prince William Sound.

"The males are roamers and singers," she told me. "You can hear their calls from miles away on a hydrophone, and they are part of the acoustic landscape of the Sound. Each group has its own language."

Originally the group she studied had twenty-two whales

but they lost over half their number after the Exxon Valdez oil spill, and now there were just seven left roaming the Sound. Eva was shown pictures of the group swimming directly through the thick of the oil. She took their extinction personally.

Now, three years later, she had temporarily left the Alaska she loved and was in exile in the place where I often wished I was. She was living with her sister's family on Cape Cod. We had begun to correspond when I was down in the Gulf during the summer. She wrote:

> If I have to be anywhere away from Alaska for six months, this does pretty well, and I'm grateful. It's very gentle here, except for the fisher killing my sister's chickens. Every day I relive the Exxon Valdez spill through news reports. It's painful. And oh so familiar and unsurprising. You may run into some of my friends in your travels there, Alaskans letting their hearts break all over again, but compelled to try and do something. Stay out of the dispersants!

In October I receive another e-mail from Eva:

> You were right; it's so much better here now that it's fall. I had a hard time with the summer crush of humanity, the Land's End catalog scenes on every beach and byway. The leaves are changing in the most subtle way. I'm used to Alaskan falls and springs . . . in and out like a lion and everything dramatic and happening so fast. I actually am homesick for that. This feels truly like an entire season. In Alaska, fall is simply a turning point between winter and summer, not a pause.

Nothing subtle about it. Leaves turn golden in a week; big wind comes; leaves hit the ground.

This is a potent time in the timeline of the spill, when the world turns away, fixates on the newest disaster, and the people in the spill region feel forgotten and lost. At least that's how it was in Alaska. People still feel that way twenty + years later.

She is right. The news has gone underground. When I try to sell an oil piece to traditional magazines, the editors snicker. No one wants to read about *that* anymore, they tell me. It's an old story. Can't you understand that we are all *sick of the oil?* Even when another rig blows up—another rig blows up!—it barely causes a ripple on the national airwaves.

I write back asking Eva for an Exxon Valdez reading list, thinking that the best way to learn what is ahead for Gulf residents is to look back at what had happened in Alaska. I want to see the future. Eva suggests a list of books that I start tackling in the evening after teaching school. Alaskan residents must have experienced a deep sense of déjà vu during the summer. As I learn more about Exxon Valdez, the similarities feel striking to the point of plagiaristic. It is all there: massive confusion about who was in charge, rage at Exxon but a willingness to take their money, a whitewashing marketing campaign to make the rest of the world forget, scientists beholden to corporate dollars, security guards and police everywhere, activists portrayed as loonies, ecosystems slathered in oil, animals killed.

In one of the books, John Keeble's *Out of the Channel,* I even read of a local fisherman, Jim Gray, who vowed to "call bullshit on Exxon's media blitz."[46] Could Ryan Lambert have read this? Is he secretly bookish?

One of the possible lessons of the Valdez spill was how

useless, and even destructive, most of the skimming and
cleaning had been—how most of it, and the specialized science, had been "false hustle," done for the appearance of
doing. Another lesson was that when in doubt, do nothing and let nature take its course. But a corporation might
understandably take away just the opposite lesson. Even
if false hustle didn't actually help ecosystems, it *appeared*
to—while placating locals by paying them to clean—and
the greatest corporate lesson of all is that appearances are
paramount.

On the other hand, every spill is different. In the Gulf,
the shoreline was not slathered and smothered to the extent
it was in Alaska. One obvious reason for this is that it occurred much farther offshore. But another was the massive
use of dispersants. Ironically, this is one of the regrets that
some people, even those of an environmental bent, have
about the Alaskan spill; a sense that if dispersants had been
used more quickly it might have saved the shorelines. If
few of the lessons from the Valdez spill were learned, that
was one that BP, at least, took to heart. Oil on the birds and
shore = huge outcry. But dispersants? People aren't quite as
eager to rail against damage they can't see.

Much of what I read reminds me of what I've already
seen in the Gulf, but some of the stories foretell the future. Lawsuits that remained entangled for years, payments
delayed, Exxon pulling out once the media turned away.
Studies revealed that for many residents the postspill
period, including attempts at reaching settlements, were
actually more stressful than the spill itself. The people I
met this summer would not have believed you if you told
them that it was actually going to get worse. But in Charles
Wohlforth's *The Fate of Nature*, I read about the protracted
legal wrangling: "Exxon kept fighting for fourteen years

more. With unlimited resources, its strategy was unlimited litigation—deposing everyone involved, seizing scientists' notebooks as they came in from the field, fighting every point on appeal to the very end."[47]

A couple of gems from the aftermath stand out. For one, Exxon's 100-million-dollar settlement fine turned out to be tax deductible. Another beauty was a court case, described by John Keeble, in which it was revealed that a group of seafood processors—called "the Seattle Seven"—were paid seventy million dollars by Exxon, but "what was not known about the deal was that the Seattle Seven had agreed secretly that if there were a later award for punitive damages they would apply for their share of it, and then, secretly, hand the money back to Exxon." Exxon wanted to *look* like they did the right thing. Keeble quotes the judge in the case against the oil company: "What is really pernicious about the Seattle Seven issue is that Exxon sought to reduce its exposure to punitive damages twice: once by informing the jury of its voluntary payment to the seafood processors, and a second time through its secret agreement with the Seattle Seven."[48]

⁂

Keeble also writes that Alaska is regarded by the rest of the country as a "resource colony." A resource colony! How perfect for the Gulf.[49]

In the summers before I traveled south to the Gulf, I traveled north to Alaska and Nova Scotia, not yet thinking of them as resource colonies. It was beauty, after all, that drew me to these places. Birds have always been my way into other worlds, worlds separate from my own, and more and more they have guided my travels: birds were what drew me to Cape Breton in Nova Scotia. In Cape Breton I was lucky enough to see gannets on their home turf and in all their glory, diving into the massive Bay of Saint Lawrence.

Gannets breed on great northern rocks off Newfoundland and in the Saint Lawrence River. Atop those rocks they barely have room to move, but what first looks like a great blob of birds is actually ordered by an elaborate system of spacing and ritual, including courteous bows and not-so-courteous bites and beautiful gestures like "sky-pointing" where a gannet turns its bill upward to indicate imminent departure to its mate. Among seabirds, gannets are the most attentive of parents, guarding and feeding their young for up to ninety days. On the other hand, when they cut the cord they *really* cut the cord. At the end of summer the young, who have never flown and hardly moved from the nest, will waddle to cliff's edge and take a great plunge into uncertainty. They will leap off the ledge and glide down to the water hundreds of feet below. Since their wings aren't strong enough to support their bodies yet, they begin their migration by *swimming* south.

On my first day in Cape Breton I watched as the wind blew a low jangling light over the white chop of the sea, spouts of water kicking straight up in the air. Dozens of gannets with six-foot wingspans gathered above the frothy ocean before pulling in their wings and dropping from the sky, hitting the water like arrows. The sea broiled as the gannets enacted a great symphonic swooping: lifting, pausing, and readjusting before again stooping down into the froth.

The next morning I watched another gannet show. Not long after dawn the birds were at it again. I returned to the same spot where twice as many gannets as the day before were diving despite the wind, which didn't surprise me. The day was raw and the white birds plunged into a green sea.

After a while I put away my binoculars and said good-bye

to the gannets. I had come north to see birds, but a friend had insisted that I also needed to visit the old coal town of Sydney Mines. I wasn't yet thinking in terms of sacrifice zones or resource colonies, but I dutifully made the long trip, driving north and then east around the entire island of Cape Breton. After six hours I finally made my way to the town of Sydney Mines, where the beauty of the rock cliffs and gannets was left far behind. The town, with its ugly brick row houses, looked like it had been cracked open and had its insides sucked out. Which was pretty much what happened.

I drove down Convent Street, the brick houses all identical, many of them unoccupied or in disrepair. Then I circled back to what passed for the town center and pulled into Mike's Place Pub & Grill. Inside the chairs and tables looked like they'd been brought over for the day from someone's home, and they were about the only decoration in the large room. A half dozen men spread themselves around the bar and the tables, focusing on their midday drinks. My friend back home had told me that Mike was a fount of local knowledge, so I asked the bartender if he was around.

"Mike's gone," she said. "He sold this place about two years ago."

I said that was too bad, that I had hoped to learn something of the town's history from him. She pointed to a man sitting at a table against the far wall.

"He's not Mike, but Keith can tell you anything you want. He's lived here all his life."

Keith was a gray-haired man with a long jaw who wore a green jacket, green baseball cap, green shirt, and gold crucifix. He was drinking his namesake, a Keith's IPA, and soon I was drinking one with him. Keith seemed more than happy, thrilled really, to regale me with the history of his hometown. Not many people ask about it, he said. He launched

right in, shaping his story as a tragedy, and therefore starting with the golden age from which Sydney Mines had fallen.
The town was once the center for the country's coal production during World War II.

"We provided coal for this country in hard times," he
said. "Back then this place, the whole downtown, was wall-
to-wall with people. It was a vibrant place, a lively town.
The people were patriotic, too, and hardworking. As a kid I
worked summers in the coal mine. We all did."

The town's history segued into his personal history for a
while, how he left Nova Scotia and went into law enforce-
ment. How his partner was shot and killed in Newfoundland.
When he, and his story, veered back to Sydney Mines it was
an entirely different place.

"The coal was gone and they had taken everything out of
the town. Where it had been wall-to-wall with people on a
Saturday night you suddenly couldn't find anyone. Maybe
a stray dog and a single taxi. It was gutted. A ghost town."

His eyes drooped as if in sympathy with the town. His
voice sounded beautiful, his accent vaguely Irish.

"There were no jobs, you see. Other than funeral directors. There was a big call for those."

This is how we do it. We hollow places out and then move
on to the next place. We move on but the people who live
there do not. Or, more accurately, they do, drifting rootless
to the new place and leaving their shell of a home behind.
Floating across the land like wraiths.

Then, with one place hollowed out, we turn to the next.

As the fall deepens, the marsh turns a golden brown and the temperature drops. The birds sweep through, some just winging past and others staying for the winter. I head to the beach to watch for the gannets' return.

Back at home I've placed a cocktail chair in the spot where I have decided to build a writing cabin. From that chair I watch the marsh and the migrating birds. It will be a flimsy sort of cabin, ramshackle and very much a shanty. It won't be built to weather a hurricane, and it will no doubt encounter hurricanes in the coming years. Ten yards from its front door is the creek that connects it to the ocean and the rest of the world.

If the cabin will be inspired in part by the night I spent at Anthony's fish camp, it will also be inspired, more book-ishly, by Henry David Thoreau. As a kid I always loved being outdoors, but I first began seriously thinking about nature after reading Thoreau's *Walden* as a teenager. The book was full of boring passages I couldn't get through, but there were also great secrets. "The life that men praise and call successful is but one kind." There was the promise of an alternate life, a life of noncomformity. And what did that life entail? Living in the woods in a small shack, off the grid as we would say today. Central to that life, of course, was the love, to the point of worship, of beautiful places.

No wonder I fought against Jim Gordon's wind farm when he first tried to place it in the waters off Cape Cod. I didn't want my sacred place to be desecrated. My eyes were not open yet to certain connections. The irony is that those connections had always been there, plain as day, right there in the book that I toted around like a Bible for all those years. Thoreau, after all, was our patron saint of fru-gality, suggesting 160 years ago that we might be happier doing with less than constantly seeking more. It was in his tiny cabin that Thoreau created the initial ledger sheet, the personal math that many of us have begun to think about again during these difficult times, the calculus of our own input and output. He did his figuring right there on the page for us. *Here is how much I spent and here is what I gained.*

Maybe, I think now, I had Thoreau's cabin all wrong. Or only partly right. It is important to remember that one of the principle motivations of his trip to the woods was economic: he needed a place to live cheaply so he could have time and space to write.

Last fall I taught Walden to a group of both grads and undergrads, and to my surprise, rather than being bored to death, they seemed to really get it. One week the homework was to not use something—a car, the computer, an iPod— and instead of whining and acting deprived, they rushed back into the classroom the next week like excited young hippies, trying to explain to me, their old teacher, that having less actually felt better than wanting more. Their eyes gleamed with conversion, which convinced me, the jaded professor, to actually go back and reread something I'd assigned, rather than merely teaching it. What struck me most on this go-round with *Walden* was how deeply economic language and strict bookkeeping pervaded Thoreau's work. It occurred to me that my own view of the cabin in the woods—as a place of solitude and wildness and romance— was a limited one. By Thoreau's reasoning, human lives, like the lives of other animals, require a strict mathematical relationship with energy, its input and output, its gains and losses, its conservation and squandering. It turns out that our lives hinge on this calculation.

On Cape Cod I was quick to embrace, and mimic, Thoreau's love of nature but slow to hear his sterner message of personal responsibility and energy use. I rationalized this by saying that I preferred Thoreau the celebrator to Thoreau the preacher. But as I, along with so many of us, have begun to wrestle with our own issues of excess and frugality, I find myself returning to the other, stricter Thoreau.

What I had missed on my earlier readings was the element of sacrifice. Thoreau's relationship with energy was simple but profound: instead of just focusing on getting more, he limited his input and refined his output. He famously argued, for instance, that by walking he could beat a man riding by train from Concord to Fitchburg, since, to be strict about it, you would need to include the amount of time the rider worked to be able to purchase his ticket. Of course Thoreau was nothing if not strict. Not many of us can boast of spending $28.12 ½ on constructing our homes or of consuming $8.74 of food in a year.[50]

What if instead of sacrificing other places—Sydney Mines and the Gulf and Alaska—and other species—killer whales and gannets and dolphins—we chose to sacrifice a little ourselves? Is that so ludicrous? Unfortunately, the word *sacrifice* has, like Sydney Mines, been hollowed out. It has become rote. Something politicians say. It has lost its heroic connotations and isn't a word people really use that much, which is understandable. Our culture has emphatically chosen the opposite route of Thoreau, focusing on getting more to the extent that the idea of *consciously* doing with less seems laughable. But what if someone came to you and whispered, "Do with a little less and two things will happen. The world will be better and you will be happier."

Well, let's think this through. It's likely that even if you bought this quiet message and tried to hold on to it, it would be swallowed up by the culture's clamor, by advertising and television and constant demands on your time from e-mails and phone calls. Even if you were a very focused person it would be natural to lose focus. Gradually you would drift back toward society's hungry message.

But it *is* possible to listen to that voice. It has been done before. Not just by Thoreau either. It's likely your great-grandparents and grandparents practiced some version of the philosophy that was so novel to my students. Maybe even your parents. It is what mature people do, after all. They sacrifice for their children. Is it possible to take this to a global level and sacrifice for *the* children?

It has been a strange fall, my body up here teaching school but my brain still mostly down in the Gulf. I am tired of watching from the sidelines and in late October I decide to follow the migrating gannets back to the Gulf.

This time I will not be traveling alone. A hairy man out of myth, a Sasquatch-like creature from the north, will accompany me. Mark Honerkamp, or Hones as his friends call him, has accompanied me on other adventures, most recently to Venezuela to follow ospreys. He is a big, bearded man, six foot four inches and closing in on 250 pounds at this point; but he's mostly jolly, and possessing enough of a sense of humor and knowledge of natural history to make me forgive him his animal ways, the flatulence and snoring and the gradual and steady devouring of every morsel of food in sight.

Long ago Hones chose a different route from his mostly college-educated friends. He took a job at the Ski Market Warehouse outside of Boston and put in his time there for close to thirty years, barely missing a day, a lifestyle that might seem unambitious to some but allowed him to fish when he wanted to and to restrict work to a box from 7:30–3:30. This made sense and worked pretty well for him until the company declared bankruptcy last February.

The man my daughter calls "Mr. Hones" arrives in Wilmington on October 24, strolling off the plane from Boston in his XXX-large Red Sox shirt. Hadley runs up to him and hugs his leg. Then I give my own hug hello, and, after a quick stop at Flaming Amy's for burritos (with extra hot sauce for Mr. Hones), we head home to huddle up and prepare for a return to the Gulf.

The next morning we fly into Pensacola and return to Fort Pickens, where my trip started this summer. On the way into the park, I point out the ghostly cleanup crews, still digging and picking three-and-a-half months later. But some things have changed: the osprey nests in the salt-dead trees are mostly empty, many of the birds having launched themselves on their annual journey to South America. One of the nests has been taken over by a pair of bald eagles, who, looking anything but fierce, lean into each other like doves.

We return to my old campsite, passing an enormous RV with the name "The Intruder" emblazoned on the side. A portable satellite dish sprouts from the Intruder's front and two plastic pink flamingos roost on the grass of the tiny front yard, signaling the owners as fellow bird-lovers.

A more natural sight awaits us at the campsite. It turns out that the monarchs have beaten us here. Shrubs bloom with butterflies. The shrubs are called baccharis, known less formally as salt bush, and their whitish flowers are now covered with wings of orange and black. Monarch butterflies float around the closest bush like a living nimbus.

If anything points to the fragility of migration, the necessity of small connections to support trips of enormous risk and distance, it is the journey of these butterflies. To think they will migrate over land and water and over *generations,* through winds and rain and death and rebirth, down to the mountains of Mexico. It sounds like a child's fairy tale, like being told that a piece of tissue will rise from your bedside Kleenex box and fly a thousand miles.

When my wife and I first moved to the South, right after Hadley was born, we had a baccharis bush outside of our apartment. I don't know that I have ever lived through a time that felt as vulnerable, as fragile, as those first years in North Carolina. As full of love as I was for my daughter, I was also full of the fear of losing her. The world felt precarious. Sometime that first September, the baccharis sprouted monarchs and I brought Hadley outside to see the dozens of butterflies. They fed off our tree for a couple of weeks before continuing their preposterous and fluttery migration south. Sometime in October the monarch tree emptied. Orange leaves fell from the bush. However, these leaves, unlike most, didn't merely fall to the ground but decamped and began a two-thousand-mile generational migration to Mexico. I remember watching individual monarchs try to fly from our beach southward over the water, dipping low, almost touching the sea, which would be the end of them, before carrying on, apparently unperturbed by their brushes with mortality.

The monarchs are the perfect greeting committee for my return to the Gulf. Welcome back, they say. Welcome back to our strange, shining, and precarious world.

V. Return

TWO STORIES

Here is the story they tell us:

Oil is over! In August a government report is released that claims that much of the oil is mysteriously gone. We are told it has in fact "evaporated."[51]

Here is what I see:

When Hones and I visit Orange Beach, Alabama, the beach really is orange. It is late October and a storm strikes during the night. Violent lightning and high surf and thunderclaps. It's the first significant rain in almost two months and the proof of it is on the beach when we walk out in the morning. The storm kicks up the oil: a brown-orange bath ring down by the shore and a line of caramel farther up the beach. It's almost beautiful in a disgusting way: as if Jackson Pollock, armed only with a candy apple machine and some cigarette butts, had a time of it making a great line of squiggling art. The line continues for miles, weaving and coiling like a python down the beach. Some discarded coffee cups are also part of the mix. It seems a great tribute to the human love of crap. And yet people here on vacation will have none of it. They stick to their illusions and walk down to the beach barefoot, their feet coated in the stuff, and lie down on beach chairs, despite the hordes of workers and the giant sand-digging machines (called "Sand Sharks") that have hurried out to clean up the new mess.

You can believe the news if you like, but all the news I

need is written in that jagged orange line on this beach. Back in Fort Pickens the tarballs were not as numerous as during the summer, but what was new was the scarp above the tide line. While summer's calm tends to smooth beaches, winter's storms sharpen and steepen them. Now a long scarp, a small cliff three feet high or so, extends along the beach just above the high-tide mark.

The scarp is a miniature Grand Canyon where you can examine the geologic layers, not of centuries, but of the last few months. In this scarp you can easily see the orange-brown line that says oil. Which points to the current problem. We are no longer talking "tarballs" so much as flat tar pancakes that stretch on and on below the beach. If the oil is visible in this cross-section, it is invisible but still present below the rest of the sandy vacation beach. In a way, the scarp sends a mixed message. It puts the oil event in the past—right there, back in July, the short-term equivalent of the Mesozoic era—something that clearly happened *back then*. You look at the rusty brown line of evidence and think "There, that is when it happened." Harder, and less pleasant, is to think that the line is still here and that in ecological terms the oil is very much still happening.

On our first day I came across a scientist named Alyssa who was supervising a cleanup crew. Instead of digging with shovels, the crew members were down on their knees, picking at the scarp line, like a squad of determined archaeologists, or maybe badgers, digging into the wall. They sifted through the sand as if panning for gold.

I asked her if there was still much oil down below the sand.

"I know there is," she said quickly. "I was here when we took core samples the other day. There are great tar mats down there. They look like vanilla-swirl ice cream."

Here is what the headlines say:

"Oil-eating microbes are helping clean the Gulf!"

These microbes are described as little "miracle-workers" that have evolved just in the nick of time—ta-da!—to clean things up. I think of Scrubbing Bubbles. The evolution of a new breed of oil-eating microbes is, we are assured, a very good thing.

BP is overjoyed with all this good news, including the government's report. They say that perhaps they have actually been *over*compensating the local fisherman. The Gulf, they tell us, should be back to normal in a year or so. Why fear what you can't see?

Here is what I see:

Hones and I pull over at a beach parking lot in Fort Pickens that appears to be a center of cleanup activity. I tell Hones to bring along the new camera and audio equipment we have been toting with us. This summer I came here more or less empty-handed, armed with only a pen, a journal, and a little tape recorder, but this time I've got bigger guns for bigger game. Hones is understandably nervous about breaking out the camera.

"I don't know, Dave," he says.

The reason for his hesitation is that the parking lot is full of milling workers, ready to call it a day and climb on the commercial coach bus—Bonanza, I think—that's there to carry them home. The workers are not roughnecks exactly, nor stevedores, but on the other hand they don't exactly look like folks you would invite to a tea party. And they don't look like they are in particularly good moods. Sure enough, when I break out the camera, it doesn't take more than a minute for two tall men, one black and one white, to hustle over to our car.

"What are you doing?" the black guy asks.

They both take a step too close to me and I go into a song

and dance about how I am here to film bird migration and of course I'm also here to highlight all the good work they are doing on the beaches. I place the camera behind me on the roof of the car, ostensibly to show I mean no harm but in fact keeping it running in hopes of chronicling our confrontation.

"You can't do that, man," the white guy says.

I smile, assuring him I won't, but my smile is not reflected back.

I put the camera in the car and Hones does the same with the audio equipment. I lock the car door. We show that our hands are empty. No guns, or cameras.

"We're going to just walk down to the beach," I say.

They escort us over to the walkway to the water, between the signs that say "Oil Spill Response Area." They stay behind in the lot once we head out over the sand.

"Holy fuck," Hones says when we are out of earshot.

"Welcome to the Gulf."

We have the beach almost to ourselves. There's only one other small group down by the water, a man from Louisiana and his family. I am still studying the scarp when the man yells out, "Jesus Christ, did you see that?"

He runs up to us and points at the water, but all we see is a retreating ripple.

"I just seen a shark chasing that school of mullet," he tells us, excited.

His accent is a thick Cajun singsong.

We search the water for sharks for a while. When we see none we walk back to the parking lot. The bus has left, taking most of the workers, including the two who confronted us, with them. I wander over to the Porta-Potty and when I get out Hones is talking to some guy in the corner of the lot. Billy Johnson—whose name I've changed to protect his job—is the foreman of the crew that just left. If the crew members were protective of the operation here, then Billy, their boss, is anything but. He has been up since before dawn and worked another long day, and after wiping his brow with a kerchief, he launches into a full-blown monologue.

"I shouldn't be talking about this stuff but I don't give a shit. I'm ready to go home. It's been five months of working sixteen hours a day seven days a week. It's been a good job and I've made good money here. But sometimes I wonder 'Are they really trying to clean this up or is this just a dog and pony show? Is this a cleanup or a cover-up?'"

I stifle my urge to run to the car for the camera and audio equipment, knowing that Billy would never say what he just said on camera. Instead I resort to more devious, Nixonian means, flicking on the microcassete recorder in my pocket.

"Let me tell you something. I supervise a crew of fifty people. But the people above me, they don't know what they're doing. Every day it's something different. One day they'll say, 'Get your shovels, boys, we're going to dig up the oil.' The next day they'll say, 'Put your shovels in the truck. We can't dig.' Then you've got these environmental advisers. They're good people; I haven't met one I didn't like. But they're a little different. I'm a hunter and fisherman. Most of these folks would have a heart attack if they stepped on a bug. We got one right now from Alaska. I said, 'You do understand that this is sand, not snow.' They'll say, 'You can't dig there or you'll change the whole beachline.' Don't they understand that it changes every day, with or without us? It's just a damn sandbar. When Ivan hit, boats blew clear through to the other side. For Christ sake's the landscape changes every minute."

As he talks I find myself thinking that Billy sounds remarkably familiar. It takes a moment to realize that his voice is almost exactly like that of the actor Slim Pickens. Now he points out to the water.

"Here's the thing. There's plenty of oil out there. I understand why they sprayed the dispersants but it will end up being worse than the oil. I say let's see what we're fighting. Let's not sink it to the ocean floor. Jesus Christ. That's the worst thing you can do. Let's get it up here so we can clean it up."

As he talks the Cajun guy from the beach walks up to us.

"I seen another shark," he says.

This leads Billy to tell us about a local incident that made the national news, when a boy was attacked by a bull shark. The boy's uncle managed to get him away from the shark. The boy's arm had been cut off but the uncle grabbed it and it was somehow reattached at the hospital.

"There're lots of sharks around here," Billy says. "But locals don't like to let people know. It's kind of like the oil in that way."

"It's all about the money," the Cajun guy sings, a one-man Greek chorus.

"The locals here drive me crazy," Billy continues. "During the summer they said, 'Don't clean it up—you'll scare away the tourists.' Now that tourist season is over they are saying, 'Please clean it up!'"

He shakes his head.

"I've taken this crap about as long as I can take it."

⌇⌇⌇

Here is their story:

Oil *is* over. Why be so stubborn, so ornery, and insist it isn't? What are you, some kind of hippie? Even the president says it's okay. Not long after the positive government report is released, President Obama takes a very public swim at Panama Beach.

Bethany Kraft calls to say that it's Bush on the aircraft carrier all over again.

"He might as well put on the flight suit and give a thumbs-up," she says. "Mission accomplished!"

But her voice is now in the minority and she knows it. She says that the spraying of dispersants by BP, whatever else it was, was a brilliant strategic move.

"Everyone gets excited and angry when they see oil reach shore," Bethany says. "But only us crazies talk about dispersants."

Here is my story:

Hones and I take a long stroll to the east, toward the rising sun. Today I, like the president, will take a swim off of a Florida beach. The water is inviting, shining green and tropical. We don't see anyone for a mile or two, but then we come upon a single man throwing a net in the water. When he pulls it out there must be twenty-five mullets in his net, good-sized ones, too. I am astounded and tell him so, but he says it's a typical cast. His name is Greg and he wears the standard beach uniform of white T-shirt, baseball cap, shorts, shades, mustache.

"See how beautiful the water is? It's the time of year when fish taste the best."

I decide not to mention oil for once. But as he untangles his net and picks out mullet, he comes around to it himself, as if he had merely forgotten the subject for a moment, a bad dream.

"Of course that shit they sprayed is in everything. And it's not as clean as it looks."

For proof he turns the pockets of his shorts inside out to show us the liners. It's where, he explains, he keeps sand fleas for bait, something he's been doing for years. Despite this, his pocket liners are usually white. Not this year. As we can see, the liners have been stained brown by the discolored fleas.

Earlier we talked with the same friendly ranger I got to know this summer.

"We just had four straight weeks of algae," she said. "Something called June grass. It is way too late in the season for that to be washing up. People were coming here on vacation and the water was like beef stew. And the June grass just stayed and stayed. Completely abnormal. People said it was because of the oil but you couldn't be sure. Whatever it was from, people left early and we took another monetary hit."

The campground stayed empty for most of the summer but there was one group of visitors who consistently rented out the camping spaces. These were people who had heard about the beach-cleaning jobs and had camped out here, putting their names on the list and waiting for an opening.

"On the other hand, the tourists who do stay are pissed that the workers are still cleaning the beach. 'What are those people doing?' they ask. 'Is that really necessary? Do people really need to use those machines at night?' They get indignant. They are ready to move on."

I decide not to mention this to Greg, and, as if by mutual agreement, we all stop talking oil and start talking fish. Hones questions him about what he catches around here. When it turns out he is having trouble untangling his net, fretting that he'll have to cut it with his knife, I volunteer Hones's services.

"My friend here is really good at untangling things. It's kind of his specialty."

Hones demurs at first, embarrassed to be put on the spot, but then bends to the task. Soon he has taken over the job and Greg stands back with me. I was not kidding around when I recommended Hones. I have seen him untangle impossible knots and know that he has fine-tuned his natural abilities by fishing obsessively over the last decade. (Once, in a terrible fishing moment, he managed to tangle a great blue heron in his line, but walked the line out to the bird, held it tight, and whispered to it while cutting the line and letting the bird free.)

Greg and I just watch as Hones twists and turns and, following a logic apparent only to him, eventually frees the net of its knots.

"Shit, I thought I was going to have to buy a whole new net," Greg says. "Thanks. I'd like to pay you in mullet."

We'd like that, we explain, but we are going to hit the road soon and will likely be spending the night in a hotel. We say good-bye and start walking back down the beach. Which is when I see a shadowy shape in the water. It is dark and large and moving fluidly underwater.

I think I know what the shadow is. Back on Cape Cod I got to know sea turtles fairly well. In the fall "our" summer turtles would begin migrating, a migration in its own way as miraculous as that of the monarchs, though this world of movement was unseen to most of us. The turtles swam south, spurred by the drop in the water's temperature, but the problem was that Cape Cod Bay, relatively shallow, would stay warmer than the ocean, and when they finally left the Bay, swimming around the Cape's hook and into the Atlantic, they were stunned by the cold water. During the winters we lived there, I patrolled the beaches and found more than a few cold-stunned turtles. Sea turtles, of course, were hit particularly hard in the Gulf this summer.

Considering this, I should give all turtles a wide berth. But I am a human, of simian roots, always curious, frequently intrusive in the affairs of other species. Before I can think to stop myself I'm pulling off my shirt and sandals, and rushing in. The water shines green, the temperature perfect. I wade out farther, to my thighs, my waist, and see the shadow just ahead. I dive down, keeping my eyes open. For a second I am close enough to touch the Kemp's ridley turtle, until, with a few quick flaps of its flippers, it shoots off. I may want contact, but the turtle is not so sure. It flies away through the water as if jet propelled.

The government report, all shiny and positive, is not the only one coming out of the Gulf. Samantha Joye, the scientist from the University of Georgia whose interview I read during the summer, has been down in the Gulf since the spill started and has published a paper concluding that dispersants and oil have been found in great quantities on the ocean floor. She has described vast plumes of oil floating beneath the surface.[52]

As for sea turtles, they have their very own report: five hundred oiled turtles have been rehabilitated since April and six hundred have been found dead.[53]

I think of the brown liner of Greg's pocket and what the turtles are swimming in.

⁓

Here is the official story:

By late August, about the time the blue-winged teal should start to arrive near Ryan Lambert's place, the national media begins to talk about the oil spill in the past tense. The news cycle is officially over. Anyone who insists it is not is looked at askance.

One night I watch Diane Sawyer interview Billy Nungesser. Nungesser's sweaty charm doesn't translate well to the small screen. He tries to explain that the Gulf isn't quite as clean as is being reported.

"Are you suggesting that BP is covering up what happened to the oil?" Sawyer asks with an arched eyebrow. *What a nut,* she all but mouths to the camera.[54]

It is starting to take on that feel. I notice how, despite the mounting evidence of great tailing dead zones, of underwater cones of oil, no one really wants to hear it. Oil is over, people, and if you don't believe it you are one of those worry-wart kooks.

The fishing grounds open first. Then Gulf shrimping season finally opens, despite the objections of many of the shrimpers themselves and reports of tarballs coming up in the nets. A cry of "Ollie, Ollie, in-come-free" goes up. Everyone back in the water.

But here is a different story:

Near the end of our week in the Gulf, Hones and I land at Ryan Lambert's lodge in Buras. Lupe is there and we hug hello. Ryan, she explains, is over at his house, about a quarter mile from the lodge itself. She immediately calls him, though I tell her not to, not wanting to bug him. She shushes me and hands me the phone.

"Come on over and we'll shoot the shit," Ryan says.

Soon we are sitting in a living room featuring the same style of overstuffed couches and high-def big-screen TV as the lodge itself. I introduce Hones and then shut up and let Ryan take over. It's not conversation we are looking for here, after all.

"Something is wrong down here," he begins. "Very wrong."

He leans back on the couch and reaches in both directions, as if putting his arms around two imaginary girlfriends.

"Where's the oil?" the papers have been asking.

As it happens, Ryan has an answer. Dig a foot down in

the sand on any of these barrier beaches and you will still find hidden tarballs.

"That's just one place the oil is," he tells me. "Another is out on the islands. Last week I kicked the white sand on the barrier islands to see tarballs below. And on the same day that the networks announced that the oil was gone, I saw slicks deep in the marshes."

He tells us that BP offered him money to run a fishing tournament to show how safe the Gulf is. The tourney would be a sign, like Obama's swim—thumbs-up!—that all is okay. "No," was Ryan's flat reply. He still hasn't gotten the money for his claim after his lodge sat empty all summer.

"It's been a strange fall," he says. "I've been fishing and hunting here for thirty years now and I've only seen one fish kill before this year with my own eyes. That was in Christmas of 1989, and it wasn't only here but all over the state. So before this summer I've never seen fish kills, and since the spill I've seen *nine* with my own eyes. Nine massive fish kills. Fish suddenly thrown up dead onshore or floating on the water. Why? For thirty years it didn't happen, not that I know of, so why'd it happen this year? Coincidence? I've seen algae bad, worse than it is this year, but I never seen fish kills. Now I'm seeing fish kills.

"Now we're here six months into the oil spill and, you know, when you talk to some of the old-timer guys they'll tell you something's not right. I'm catching speckled trout right now. Usually in October, when the trout come in, you have ten to twelve boats out fishing which means you're catching a thousand fish a day. But that's not what we're see-ing. I've seen only seven boats reach their legal limit since July. Seven boats! Unheard of. Ought to be seven a day. I can understand why we don't have business because of the perception of the oil. But not to have fish. And not to have the birds working in the barrier islands that usually are just loaded with birds. I went out here the other day and there's not a gull. There's not a tern. There's not a roseate spoonbill or egret or crane. Pelicans, nothing. They're completely barren. Why?"

We sit back and listen. I don't say anything other than a few "uh-huhs." We are in Ryan's world now. And if one voice has to narrate the spill for me, it might as well be his.

"The thing is, there's still a lot of oil coming in. You wouldn't know it from the news, but there is. What's today? October 28? Well, from October 1 to October 11 we picked up 36,000 gallons of oil just in Bay Jimmy, to the west of here, and 10,000 bags of tarballs in those eleven days alone. So it's still bad, it's still coming. We're having all these algae blooms, too. This morning I read a report that said when you emulsify the oil in the hydrocarbons it makes it more conducive for algae blooms and we'll have more and stronger algae blooms in the future and we can expect more fish kills. Because of the fucking hydrocarbons."

"They just had a huge algae bloom for three weeks in Fort Pickens," I tell him.

He nods.

"Down in Venice it's really bad this week and it's all algae. Only time will tell, but as a self-proclaimed expert on the outdoors in this area? I'm not a scientist, and I won't be able to put my finger on it and blame a certain thing. But I'm telling you, it's going to be a long time getting better. Now I'm seeing fish kills. Now I don't see speckled trout. Now I don't see birds. Something is very, very wrong."

He stops. His face is red. Maybe it's just another day out in the sun. But if I didn't know better, I would say that the big man in front of me was about to cry.

The next day we again stop by Ryan's lodge. The place is

empty except for Lupe in the kitchen and Ryan, at the table, spooning up a bowl of duck gumbo. He insists we join him and when I offer up some fairly feeble demurring—"No, no, we couldn't"—Hones looks at me as if I am insane. Soon we are happily shoveling the gumbo into our mouths.

"Yesterday we were cleaning shrimp," Ryan says between bites. "We had five hundred pounds of shrimp and I looked at the gills and I could see black inside there. It's just not right; I don't know what that black is but it's not right. Sure enough I get a call this morning from another experienced guide and he says, 'Yeah, I was cleaning a hundred pounds of shrimp yesterday and I think I'm hypersensitive but now the—' And I said, 'Stop right there. Let me tell you what happened. The gills were black.' He said, 'That's right, how did you know?' I said, 'It's not you and it's not me, and I'm glad we're not paranoid.' Because you know I try to be open-minded about all this to make sure that I don't overstep and accuse somebody of something they didn't do. But things are not right. I know when things are right because I been here so long and I live outside. I know when it's going to rain; I know when it's not; I know almost everything about what happens outside and it's just not right."

When we finish, Ryan leads us out back to the fish-cleaning tables.

"I'll show you what I was talking about," he says.

We walk out back to the shed where the trucks and boats are kept, and where a mountain of shrimp rises off of blood-stained tables that sometimes see a thousand fish a day. One of Ryan's workers, whom he has brought up from his lodge in Mexico, is sorting the shrimp and cutting their heads off, and Ryan starts to speak to him in Spanish. Ryan picks up a few shrimp and flips them backhanded into the pile before finding one to his liking. He holds it up in front of our eyes.

"Lookit that."

We do and see a fat shrimp in its semitranslucent shell. It looks tasty. Except for the dark black stain where the gill is. Ryan picks up another and we see the same stain.

"When we first noticed this we reported the black on the shrimp and they said, 'Well, yeah, that's black gill disease. It's a bacteria.' So okay, I'll buy that, a bacteria. So then I got a question for you. Why haven't I ever seen it before? I mean I've been here thirty years and never seen it. Why did we bring in five hundred pounds the other day and almost all of them have it? Now today we got a hundred pounds and not all of them have it, but as you can see there are some in here. But yesterday it was just everywhere, and why?"

We have no answers for him.

I have been reading articles and listening to reports broadcast from the Gulf all fall. But this, it occurs to me, is a different sort of news.

FAITH

But I am getting ahead of myself. Let's back up a little. On our third day in the Gulf, Hones and I drive down from Fort Pickens to Orange Beach to the west end of the peninsula. We make note of the bars we pass. Tippy's Tavern with a picture of a tilted beer mug with foam spilling off the top. Panama Mack's Bar and Restaurant. Rocky's Bar, with an impressive daily happy hour of twelve to six o'clock.

As for the nonalcoholic landscape, it is low and green and storm vulnerable, full of marshes and herons and egrets, reminding me of coastal North Carolina. Though I have resided there for almost the entire seven years my daughter has been alive, I still have no illusions that I am "from" Carolina. But the South is my adopted home, and I have taken some pains to learn my place. While I may never completely embrace the culture, I have begun to experience a sort of deep attraction and affection, something that I am not quite ready to call "love," for the land. It is a low land bursting with birds and the fish that birds often feed on, a land of carnivorous plants and twisting junipers and tall pines and shaggy moss, a land packed with and protected by barrier islands that are always in movement, dancing with the many storms. In fact, if you follow the coast from where I live to where I am right now, you will find more barrier islands than anywhere else in the world. The islands, of course, are the jewels of the coast.

Yesterday we drove into Gulf Shores and then doubled back to Orange Beach, coming upon a stretch of enormous hotels that line the beach. They looked like blocky castles in the mist and as we passed by I said their names out loud: The Shearwater, Windward Pointe (with that fancy final "*e*"), Pelican Pointe (again the *e*), The Sands (of course), Blue Water (not anymore), and Summer House (for a giant). After I passed the last one, I banged a U-ey, deciding to spend the night in the smallest of the giants, a pink-walled Holiday Inn Express where I'd stayed this summer.

I was here during the worst of times, right in the summer thick of the oil, but for the two days I was in town the mood was not particularly grim. In fact, if anything Gulf Shores and Orange Beach seemed festive, people bubbling with excitement because the word was that *he* was coming back. In tough times people look for saviors, and there was a fervor in the air as they anticipated his return. While it was true I was in the deep South, it was not Jesus of whom they spoke. Rather it was a balding entrepreneur named Jimmy, who for many years had denounced this, his own true birthplace, while fashioning an origin story around Key West. They say it was his sister, who ran a hotel named after him, who twisted his arm and got him to come back. They say it was his sister who set up the great oil spill benefit concert on the beach. Whatever the origins, the great Jimmy Buffet was back.

By sheer coincidence I had pulled on a tie-dye shirt the morning before the concert, and due to this, I was the subject of ornithological misidentification throughout the day, everyone assuming I was a Parrothead. I denied it as best I could, but what was the point?

After throwing our bags on our beds, Hones and I headed over to a nearby restaurant called Live Bait. Hones has now made it his personal mission to sample as much

Gulf seafood as humanly possible. On the way out of Fort
Pickens we ate at a restaurant named Pegleg Pete's, where he
devoured a plate of fried oysters, and at Live Bait he continued his diligent sampling of Gulf seafood, putting away a
mixed platter of shrimp and oysters. I suggested to him that
after the trip he should donate his body to science so they
could study the deeper effects of the spill. He responded by
asking if I was done with my food and then reaching for it.

It is always about now, day three, when we start to drive
each other crazy. I have no problem with the eating, but the
snoring is another story. At Fort Pickens it was a good thing
we had chosen an unpopulated corner of the campground
because I was pretty sure Hones's snoring could be heard
for miles. I was in a separate tent but still had a hard time
falling asleep. The hotel, of course, was worse. In the same
room and not separate tents, it felt as loud as being trapped
in a phone booth with a howler monkey.

Traveling with Hones is like spending the holidays with
your family. Enough time passes between the last time for
you to conveniently forget what it is about these people that
makes you want to kill them. With Hones it's not just the
snoring but the backseat driving and the checking that the
doors of the car are locked, as if, without his guidance, I
would not have a chance of navigating through the world.
Yet I understand that this last is a product of his living—
and usually driving—alone. Also, I know that from his point
of view, I must seem like a little dictator: let's go *here,* film
that, interview *him.* But we are friends—old friends, good
friends—and we can laugh at pretty much anything, our own
faults most obviously and most often. So, overall, we do okay.

Today we take the ferry to Dauphin Island, through a
shallow Mobile Bay full of natural-gas derricks and pelicans.
I give Hones a quick tour of the miraculous termite sand

mounds of Dauphin Island—still there all these months later—and we talk to a pretty young mom, her two little girls with her, who tells us that she has just had her family's home moved inland. Her house is now high up on stilts, twenty feet above the ground, smack-dab in the middle of the island.

"I like it that way," she says. "I feel safer. We decided to buy this property and move the house here from the beach side. The house had started falling in the water this September. We knew we were in trouble when the tide came in the front door."

From Dauphin we head to Bayou La Batre, which is barely recognizable. No more hustle and bustle, no more Vessels of Opportunity, no more paranoid police state. Just a patch of dirt off the Bayou where a couple of men sit in lawn chairs and lazily toss lines in the water. The whole cleanup operation is gone—*poof*—like the point in a conspiracy thriller when the duped character swears to the authorities that there used to be a business here, right here, in this vacant storefront.

We sit on the dock and watch the bayou. A dolphin leaps out of the water, its slick back shining. A wharf rat scurries under the pier. Only one boat sits in the water at the launch site, a whaler about fifteen feet long. A mustached man in a cowboy hat stands in it, fiddling with some boat lines. I wander up to him and ask if he might taxi us out on the bay for fifty bucks. Not exactly the kind of money BP offered this summer, when boats like his got fifteen-hundred dollars a day as backups, paid for sitting still and being "on reserve," but a reasonable enough figure, I think, in this quieter climate.

"I'd love to, my friend," he says. "But my crew's on the way. We are doing some restoration work across the water there."

He points and squints, and I think the cowboy hat is not such a bad fit. We start talking and I find out his name is Jim Duffy and he has plenty to say. He may look like a cowboy but he talks like an evolutionary biologist, which it turns out he is, having gotten his degree from Mississippi State.

I mention how quiet things are and how busy they were this summer.

"Well, they might have looked busy but that doesn't mean they *were* busy," he says. "About the same thing is being achieved right now as was then. This is the normal bayou and the normal bayou pace."

I ask him about the current mood in Bayou La Batre.

"The mood? Oh, doom and gloom. First, the decline of the fisheries, then Katrina where half these boats ended up in the woods, then the drought and the influx of the oyster drill, which preys on the oysters, and now the oil. Most of the people are shrimpers and they have it pretty bad. First of all they have no control at all over the price of their product. By the time I started working for the conservation department, back in 1992, already 85 percent of the shrimp consumed in this country was farmed and non-native. These guys have to compete with foreign farmed imports, too. And this is before you add oil to the mix."

I nod. It's a truly gloomy picture.

"Is there any hope at all?" I ask.

He thinks about it for a minute. He moves to the bow of his boat and stands a few feet from where I am on shore.

"Yeah, there's hope. There's ship-building here, which is a kind of silver bullet. And we have a sustainable mullet fishery right out there."

He points out to the mouth of Mobile Bay and I think of the net full of mullet that Greg got yesterday with a single cast.

"Right here we are at the edge of an ecosystem shift. If you go that way, to the west, the water becomes muddy from the Mississippi and the mullet tastes muddy too. But from here to Florida the water is clearer and the mullet tastes great. And because it's a tightly schooled fish and schools by size, fishermen can fish all day and just catch mullet, without a lot of bycatch. There's a demand for it, too. So everything about the fishery speaks of sustainability. The catch is clean . . ."

But that is just one fishery, he admits. He believes that, overall, fishermen must adapt and, as hard as it may be to swallow, find something new.

"I'm not sure the oil is going to do anything but help in that way. Of course it's a disaster. But it is going to keep putting a lot of money into the local economy. BP money, research money, relief money. If people use it intelligently then there's a chance they can transition from fishing to something else."

Then he shakes his head, as if afraid of getting carried away.

"You asked about hope. Well, there's hope that has nothing to do with the real state of the fisheries. Fishing communities in general, and this one in particular, tend to be faith-based. In hard times there's always been God. Fishing people rely on him for everything. They depend on him for the creatures in the water they catch. For the weather. For their health and well-being when they are out on the water. They pray constantly.

"What I'm getting at is that the people who really trust in God are the ones holding out," Jim continues. "They know all the rational stuff about the number of fish and they know about biological issues and depleted stock. But God is on their side and they are holding out for the miracle. While most of us see a future of slow decline and steady regression, they are holding out for the fish to come back and the regulation to go away. For it to be like old times. They don't believe in the modern story just like they don't believe in geologic time."

I had assumed that Jim was speaking as a scientific observer, an outsider, but that is not the case.

"I'm a very faithful person so I struggle with this all the time. I know that the planet is 4.5 billion years old but I also have a book I believe in that tells me man was created ten thousand years ago when Adam was born. I believe in what the rocks tell us, but I also believe in God. The rocks tell one story but there are other stories. I'm not even conflicted about it anymore. At some point I just decided it wasn't something I was going to worry about."

It seems he is done and for a minute his face is hidden by his hat. But then he raises his head, as if to scan the lot for his tardy crew.

"I don't think anybody *blames* God though. People who live here know that living by the sea has always been a difficult proposition. The oil may be new but it's always been something. Hurricanes, for instance. They're just the price you pay for living on the coast. It's the dynamic interface between the land and the sea. The interface is a tremendous area of change. And danger."

We talk for a while more until his crew finally pulls up. They are a cultural mix worthy of the *Pequod*: Cajun, Vietnamese, Mexican. They cram onto the small boat. Before I go Jim wants me to meet one member of the crew, a wildly bearded man with suspect dental work, born and raised in Tennessee. He has spent his whole life working with native plants, traveling all over the country like Johnny Appleseed. The man is shy and will barely talk to me, but I think I understand why

Jim has insisted we meet. Here, he is all but saying, is another option. Here is someone who, rather than fishing the empty seas, now spends his life regrowing the shore.

As we drive away I think about faith. I don't have a ton of it, at least not of Jim's variety. And I'm not sure I believe in our ability to change for nobler reasons. But we may be forced to change, when push comes to shove. The question is, what will be left when the pushing starts? What damage will we have done? What will be irretrievable? What forever lost?

⁓

We speed north to a meeting of the Alabama Coastal Recovery Commission that is taking place in a fancy hotel in downtown Mobile. Bethany is there, dressed as nicely as her surroundings, and all the men wear suits. Hones and I, meanwhile, sport T-shirts and cargo shorts, muck from Bayou La Batre still on our shoes. At one point, as we are taking our seats, we cross in front of the projector and Hones' shaggy silhouette is thrown up on the screen, there for hundreds of well-heeled Alabamans to titter at.

"Damn you," he says.

He is more serious than joking. One of the reasons he took his job working in the ski warehouse was so that he never had to dress up for work. And now we, still smelling of the bayou, are in a room full of suits. I calm him by promising a big meal and many drinks after the meeting, and we settle into our chairs and listen to various politicians, newspaper publishers, and scientists tell the room how they are going to lead Alabama back from the brink. The oil, they say with glossy PR smiles, is really an opportunity! If the delivery is glib, the message is not so different from Jim Duffy's a couple of hours ago.

"Alabama sure got screwed when it comes to the coast," Hones mutters when they flash a map of the state on the big screen. It's true. By any geographic logic, much of Florida's panhandle should have been theirs, and what they were left with was essentially the small stretch that we traveled this morning, from Orange Beach to Dauphin Island and Bayou La Batre. But one of the lessons of the spill is just how much that slim slice of coast provides for the rest of the state, not just seafood, tourist revenues, and taxes, but culture and nature. For all the stiffness of the present company, it is a lesson they seemed to have learned, or at least are repeating as if they've learned, and they are determined to use the pot of expected BP money—money that will be paid to the state for damages—not just to fix what was, but to bolster and regrow the coasts.

"The bullet missed us," the man at the podium says. "We didn't dodge the bullet, but it missed us. . . ."

People murmur as if this makes sense, though I'm not sure what the difference is as long as the bullet didn't hit them. Then a cartoon is projected up on the screen. In it a man is sunbathing at the beach and looking out at a pretty blue Gulf that is labeled "Reality" while a black hammer called "Perception" rises out of the sand behind him, ready to come crashing down on his head. Well, yes, sort of. Though if I were to walk up to the screen with a Magic Marker I might draw a black cloud called "Dispersants" under the water's placid surface. One of the goals here today, expressed by a local seafood magnate, is to launch a campaign to convince the rest of the country that the region's seafood is safe. Which is fine—as long as it is.

While the messengers might be cheesy, there are parts of their message I buy. The more I travel, the more I see how all coastal issues are entangled. What is happening down here is about so much more than oil. At least part of it is about how humans have decided to treat, and live with, the shore. These people seem to get that. It has been a wake-up call of sorts: when oil threatened the coast it showed the value of what was threatened. Not just tourism and fish, but culture, identity, and delight.

There is much mulling afterward, and coffee and cookies. I shake hands with George Crozier, a friend of Orrin Pilkey's who heads up the Dauphin Island Sea Lab. There are more hands to shake, more schmoozing to be done, contacts to be made, but Hones is sick of it. When Bethany tells us she is off to meet with the subcommittee on environmental matters, he starts edging toward the door. He tugs my sleeve as if ready to pull me with him.

"You promised," he mutters.

So we head out into the streets of downtown Mobile and before long we have settled in environs more to Hones's liking. The Bicycle Shop on Dauphin Street is a brick building with an open-air bar out back and a menu full of local favorites. He drinks rum while I have a martini. It's not yet three thirty but already I sense a big night coming on. Too quickly we order a second round from Anya, our beautiful Russian bartender.

I start up a conversation with our neighbor two bar stools down. He's an overweight man with suspenders who calls Anya "sweetheart" when he asks for another drink. When I tell him I'm here to cover the spill he laughs and shakes his head.

"The fucking media," he says. "No offense."

"None taken."

"All summer long the same damn picture of the same damn bird," he continues. "I hope that bird's living a good life somewhere in a cage with someone spoonfeeding him. He deserves it. After this summer he must be tuckered out."

He takes a slug of scotch.

"The media overreported it this summer. On the other hand, they're underreporting it now ... on the other hand ... well, the whole fucking thing is other hands."

I know what he means. Did they dodge a bullet? Or is the dispersant another type of bullet? On the one hand it's not as bad as some people thought, but on the other it's worse than they are saying now. He's right—it's all other hands. It kind of makes your head spin.

We are joined by Sergio, a friend of Bethany's who works for Fish and Wildlife surveying the oil damage. He can't talk about the surveys, he tells us, but he can talk about his two great passions: beer and birds.

We switch to drinking a local microbrew that Sergio recommends, and he tells us stories of the early cleanup, when he was stationed as an advisor up on the Bon Secour beaches, near where we climbed on the ferry this morning.

His are tales of wonder. Legends of the spill. He worked nights and found it much more relaxed, and infinitely more surreal, than the day shifts. The cleanup crews all wore headlamps, and they put red tape over them, and over the ATV headlights, so as not to scare away the sea turtles. It would be pitch black and you would see disembodied red balls bobbing up and down the beach.

"One weird thing was how much power we used," he says. "Sometimes they would run a generator that would flood the beach with light for our midnight lunch breaks. And the coach buses that brought the workers would run all night with their air conditioners on.

"But I saw things I'll never forget. I remember one night when the whole scene was lit up by a full moon. And another when the phosphorescence turned the ocean a glowing green. I heard that workers and supervisors sometimes fought during the day shifts, but we all got along at night. The

contractors were great for the most part. And the cleanup
crews seemed to actually like what they were doing and to feel
it was something important. A lot of them were kids pulled
off the unemployment lists in Pensacola and they were suddenly making eighteen dollars an hour, more than they had
ever made in their lives. People wanted these jobs so much
that some guys showed up and started to work without being
paid, in hopes of being hired. I started to get to know some
of the workers pretty well and a lot of them were proud of
the work they were doing. The problem was that by working
hard they eventually eliminated their own jobs. I felt terrible
when they got laid off once the beach got cleaned. You almost
found yourself thinking, 'Oh no, there's no more oil.'"

There were rumors that angry locals were digging holes
and running chicken wire across the beach to mess up the
ATVs. But Sergio didn't see any of that. The only problem
was the local cops who would come screaming through the
high dunes on their own ATVs. "We want to make sure
there are no fights," they said, but they were the ones itching
to fight. It was park service jurisdiction, not theirs, but they
couldn't help themselves. They needed to flex their muscles.

"Maybe they just didn't like having black people on their
beaches," I suggest.

"You said it, not me," he says.

After a while we return to Sergio's house, where Bethany
meets us. Sergio's backyard is a strange southern jungle with
a fridge holding home brew and good out-of-the-way spots
to urinate. We stay late and talk long.

⁓

The next morning Bethany drives Hones and me back to
Bayou La Batre.

She talks about her own reaction to yesterday's meeting.

"It may have been a little cheesy," she admits. "But at least people are talking about the coasts. The point is that the oil spill didn't push the coasts to the edge. We were already hanging precipitously from a cliff. The oil spill just points that out. If there is one good thing about the spill it is that we now have one last opportunity to improve how we live on the shore. I hope we can do it."

The project we are going to see is based on that hope. At the landing in Bayou La Batre we are greeted by Jeff, who works for the Nature Conservancy and has supervised the installation of a mile's worth of oyster reefs. He pulls his boat up to our dock and we climb in and head out toward the reef. He is a big man with a red beard and looks more lumberjack than sea captain, but he handles the boat expertly as we cruise out into the mouth of Mobile Bay. Bethany explains that the goal is to eventually place one hundred miles of reef out along the Gulf, and that the Nature Conservancy's project, paid for with stimulus money, is the first step, and template, for that goal.

We pull up to the backside of a small island where an egret hunts in the beach grass. Jeff points down at the first reef, a creative combination of human and oyster ingenuity. Oysters can produce over a hundred million eggs but the larvae need something to attach to, and that is where these reefs come into play. Half of the Gulf oyster beds have been lost in the last few years, but these are still the most productive oyster grounds left in the world. The forces working against the oysters include increased salinity in the bay, drought, the predatory oyster drill that, as its name suggests, drills down into the shells, and, now, oil. The reefs are an attempt to stack the cards back in the oysters' favor.

Jeff jumps from the boat to the reef, this one made out

of mesh bags full of oyster shells. Water sloshes against the oysters. I climb on the reef, too, and stand there atop the shells, twenty yards behind the backside of the island in shallow water.

These shells, I know, come from the great mountain of shells that we saw on the drive in. I first saw the mountain this summer and was confused. About halfway up the dirt bayou road a huge midden of oyster shells rose from the ground. The pile rose thirty, forty, fifty feet high, and I pulled the car over at its base. What did the mountain of shells mean? Had some innovative captain sent out shell-less oysters to restaurants? Had a particularly hungry crew of fishermen polished off a few hundred thousand oysters? Now I have my answer. The pile was destined for these reefs and the place they were piled was the headquarters for the Conservancy's contractor, Wayne Eldridge, whose local company is building the reefs.

Jeff climbs down into the water to demonstrate the hard work of hefting the bags into place. After a while we get back in the boat and putter over to the next reef, which is made of triangular boxes full of oyster shells and to the next, which is made of what look like concrete diving bells filled with shells.

What exactly are these oyster reefs good for?

Quite a lot, is the answer. Most obviously they are good for oysters but that, it turns out, is just the beginning. They also, in no particular order, provide a habitat for fish and hundreds of other creatures, filter and clean the water (each oyster filtering up to twenty gallons a day), battle erosion on the local islands, help grow sea grass, provide an alternative to groins and walls, and protect the mainland from hurricanes and oil.

Oh, and they also grow fast.

"Once the reefs are placed in the water," Jeff says, "the growth is almost immediate. Young oysters cling to them. You can see the sediment being trapped right away. Which takes sediment out of the water and leads to the growth of sea grass. Which in turn anchors the island. After a while the marsh grass will migrate out and join the reef and in ten years we can put another line of oysters in. All the while you have growing islands and harvestable oysters again."

Then there is the fact that Wayne Eldridge employs local workers and that this gives the residents of Bayou La Batre a new way to work out on the sea. All and all, a pretty big contribution for a few thousand bags of shells. Yesterday Jim Duffy somewhat vaguely suggested that fishermen would need to "transition from fishing to something else." But here the vague becomes concrete. The effort to protect the shore becomes a new way to employ those who live by it.

"This is just a mile of reef," Bethany says. "Imagine a hundred miles."

It is a dream shared by her organization, the Alabama Coastal Federation, and the Ocean Foundation (where Bill Finch works), as well as the Nature Conservancy.

From the reefs we head out to the far west end of Dauphin Island, a privately owned, undeveloped section of sand and beach grass that was cut off from the rest of Dauphin during Katrina. We anchor on the bay side but soon are marching across the island toward the Gulf, drawn by the surf's roar. We find the usual—tarballs and an orange-stained shoreline—but also the exotic: about a half dozen dead man-o'-wars, bright blue creatures that are deadly but look like plastic toys. I walk up the beach past more man-o'-wars and whelks and scurrying ghost crabs. Also dozens of migrating plovers pecking on the sand flats. From the east end of the island I can see the mile-and-a-half-long riprap groin that now connects the sandbar I'm standing on to the rest of Dauphin Island. This was another of Dauphin's emergency measures, nominally to stop the oil from flowing through the cut (though it was just completed about two weeks ago, so you have to wonder how it could have done that). The project cost fourteen million and, according to the Army Corps of Engineering's emergency permit, will be allowed to exist for one year. That means that, in theory, they will be taking down this just-completed project next June. The double row of rocks was built, not just to hold back the oil, but to fight the increased salinity of the water. While these outcomes seem unlikely, one real result is the erosion that strafes the western end of the island.

How much better would it be to use that fourteen million on oyster reefs? But I can't have things both ways and there is no denying that the groin project is a boost to the local economy. As usual, it's complicated. When I mention this to Jeff he tells me that it's even more complicated than I think: the contractor who is building the groin is Wayne Eldridge, the same man who's building his reefs. This fact doesn't bother me, though, not really. Orrin Pilkey might declaim against the groins and I am with him. But I am also on the side of messiness, and while Mr. Eldridge's motives may not be entirely pure, I am encouraged that he, when not working against the environment, is working with it, building oyster reefs for profit.

We climb in the boat and soon we're flying across the water. On our way back into Bayou La Batre, Jeff has a treat in store for us. We pull up to Cat Island—really just a sandbar with a few humps of beach grass—and hundreds of birds take flight. It's joyous enough to see the brown pelicans and cormorants lift off, but then comes the real delight. Giant birds—white and radiant—fly beside the other smaller species, dwarfing them. I have never seen them before but right away I know what they are. They live in the West and have migrated down here, as they do each winter, through the Rocky Mountains. With nine-foot wingspans they lift off like bulky angels, their white wings marked with vivid black outlining streaks. As they fly off in front of the boat, Hones and I yell to each other. *White pelicans!*

Earlier I questioned the wisdom of spending the day looking at some oyster reefs. Now I'm thankful we came along. Traveling through the land of tarballs your mind can grow dark. But here at last is something to have faith in, something that fills me with hope despite the larger hopelessness. First the oysters, cause for a practical sort of faith. I believe in the reefs and I believe in the people, like Jeff and Bethany, who are doing this sort of work. Use nature as your ally, work with it instead of imposing on it, and the results might surprise you. Like Ryan and the Mississippi, or Jim Gordon and the wind, they are scheming for ways to put nature to work. This is not always a pure endeavor—what is?—and both motives and results are mixed. But at least they are doing something, taking action while others sit on the couch and despair.

But while I believe in what they are doing, that belief is too rational to go by the name of "faith." It's not the same as what I'm feeling when I see the white pelicans. What I experience when I see the birds, these great white radiant birds, is more akin to what Jim Duffy described when he said he could believe in both a certain book and the rocks, God and science, even though they tell different stories. Like Jim, I can believe in two stories: the pessimistic eco-story of my tribe, and, at the same time, a greater, wilder story. That story has nothing to do with words or the future or how we will or won't act. It is happening right now. It is an irrational story, an ineffable one. It is about the birds themselves. It *is* the birds themselves. White. Radiant. Flying.

I am not a religious man. But as I watch one white pelican veer away from the rest, my body fills with something that I have no words for. I don't have an organized system of belief. But I do have faith in that single white bird.

What is faith if not belief without, or beyond, reason? That is what I have in nature, even at this late date in its destruction and demise. I understand that we are at the end of nature, that it is dead and outdated, and that I'm kind of old-fashioned for believing. But still. To say it is as close as one can get to going to church has become cliché, but being out here with these birds does offer me at least some of the pleasures and consolations of religion. It offers me a place outside of myself, a place to consider things beyond me, a place of wonder and awe. It is where religions were born. Couldn't the first primitive imaginings of angels have been sparked by the sight of white pelicans?

Ethereal thoughts, but I bring them up for a practical reason. BP has just quietly announced that it had a 1.79-billion-dollar profit in the fall quarter. Good for them. Now I want to declare my own profits and losses. If I am going to keep an honest ledger sheet, and tally up our deficits and gains, then these spiritual issues must be given some account in the tallying. Because along with wetlands we are losing this: a place other than human, a place not smeared with our clumsy thumbprints, and a place, since we are being practical here, with the distinctly human use of seeing beyond ourselves. It seems reasonable to point out that for some of us BP has soiled not just our beaches but our church. Isn't there room for that in an honest tallying?

The engineers and accountants might not put much stock in my report, but in this uncertain place in this uncertain time there is one thing I can say with confidence. It is a truly miraculous world we are destroying. A world where shrews can somehow become dolphins. Think of that. Think of the delightful fluidity, the sheer thrill of adaptation. Could straight lines lead to this, could engineers plan out how to get from the A of shrew to the B of dolphin? "Miraculous" may have strictly religious connotations for some, but I'll stick with it in this instance. You can believe that this is God's creation or you can believe we evolved. You can even believe both. That is not my fight at the moment. But whatever your beliefs, and whatever your origin story of favor, how can you not believe in dolphins and white pelicans?

The old lesson that we taught our children was the sheer wonder of the great interconnecting web of life. But who has time for wonder when dealing with the next emergency? Though most of us dimly perceive that the world is connected in ways we don't understand, we are too busy to *oohh* and *ahhh* when the white pelicans fly overhead or the dolphins leap or the cedar waxwings fly into town to eat the winter berries. But if the world can no longer teach us its lesson through childlike wonder, it has sterner means at its disposal. After all, you don't have to understand the latticework of connections to feel the results of the latticework being broken. That is the thing about a web. It takes the genius of time to weave it, but, as hard as it is to construct, it's easy to rip apart.

ENERGY

Lofty thoughts give way to base needs. We are tired and we are hungry. When we get off the water we decide to have lunch at a tiny Vietnamese restaurant called Pho. There is a substantial Vietnamese fishing community in Bayou La Batre, and this is where they come for noodles. The Vietnamese owners have seen a lot in their day but I am confident that they have never seen anyone put as much hot sauce on their noodles as Hones does now. When the broth looks bloody enough, he goes to work slicing up little peppers and plopping them in. Hones is a pepper freak and during the trips he's often fretted about the pepper plants he left untended in his apartment back home.

When I don't finish my beef and broccoli, he takes it and drowns it in hot sauce. When Bethany doesn't finish hers, he asks if he can polish that off, too.

"I don't like wasting food," he says by way of explanation.

"You sure don't," I say.

He ignores me and reaches for Bethany's bowl.

"I'll give it a good home," he promises, rubbing his belly.

After lunch we thank Jeff and hug Bethany good-bye. She is going back to Mobile and we are heading on to Louisiana. I nap through much of Mississippi while Hones drives. He gets us past New Orleans and then, on the way down to Buras, we stop for beer and gas and I take the wheel. We pass signs that say "Plaquemines Parish Thanks Billy Nungesser" and billboards that ask "Oil Spill Claim?" and advertise for lawyers. This is Hones's first time here, and as we descend into the sunken land he takes it all in. He points out birds and another sign that reads "Elevation 4 Feet." I'm actually surprised that those numbers are positive. If most of the world's scientists are right, this will all be underwater by the time my daughter is my age. But no one quite believes it— people *live* here, after all.

It's dark by the time we reach Buras. We head, not to Ryan's lodge, but to the environmental organization's headquarters, where we will spend the next couple of days. Before we came down the organization offered to pay Hones for his photographs and videos, which I thought a coup for a guy who had recently lost his job. But he declined, worried that it might somehow disturb his unemployment checks. He did agree to sign a release, however, so we now have a place to sleep. I am particularly happy with the fact that we can take bunks, and rooms, on opposite ends of the hallway, effectively ending the snoring crisis.

After we've thrown our things in our rooms, we head over to the lodge, where Ryan invites us to his house and announces that "something is not right" out on the water. We listen for a long time to his grim message. More happily he also tells us, as we are leaving, about a place to fish near the lodge. It's late by the time we get back to the headquarters. We drink beers and turn on the TV and watch some Time Life infomercial about singer songwriters from the seventies. In between snippets of John Denver and Cat Stevens, we talk about Ryan. For his part, Hones can't believe he was in the same room with the guy who lives out his own dream job. It took about a minute for him to be completely in Ryan's thrall.

"That guy's fucking amazing. He really *knows* his place."

A high compliment in Hones's book.

We sleep well and the next morning we drive down to Venice. Not long after Halliburton Road, I pull over and take out my binoculars. Just like this summer, there are plenty of egrets and herons and cormorants. A black rail runs across the road, a sneaky dark-cloaked spy caught momentarily in the open. Hones can see the fish jumping and is tempted to take out his rod, but decides to wait until the afternoon. I notice that there are many more ospreys than there were this summer. For some this must be their final migratory destination. To the south, out by open water, I see some large birds that I think might be gannets, but then again that might be wishful thinking.

What we do see, right before we leave, are a half dozen wheeling white birds that gradually clarify themselves into pelicans. For the second time in two days I am being treated to the sight of white pelicans, this group of five soaring upward in great swooping spirals.

Their grace is impressive, but so is their sheer size. To put this in perspective, consider that an osprey, a bird that many nonbirders confuse with eagles, usually weighs about four pounds. People can't believe it when you tell them, but ospreys, like all birds, weigh next to nothing due to their hollow bones. Then consider that brown pelicans, generally considered a *huge* bird, weigh twelve pounds. Finally, with that as context, turn to the white pelicans that we now watch as they, seemingly weightless, spiral upward. Their weight? Twenty-five pounds.

I remember something that the guy in the bar in Mobile said.

"Brown pelicans are tough birds. But one white pelican can kick the asses of two or three brown ones."

The birds vanish, becoming white specks, then nothing.

We climb back in the car and drive north to Buras. Ryan had mentioned that he might take us out on the water, but when we stop by the lodge he says it's too windy. The weather has shifted dramatically. This morning we were sweating, Hones especially, but around noon a cold front swooped in and the wind came down from the north. The fishermen staying at Ryan's lodge are disappointed, but Hones is relieved. He hates the heat. In the late afternoon we pull into one of the few restaurants in Buras, a drive-through daiquiri bar. That it's a strange operation goes without saying, but the bartender tells us that daiquiri bars are common in Louisiana. We sit on stools at the bar, while over in the corner men and women play video slot machines. I order a margarita and Hones a piña colada, and we sip them while the bartender, who grew up here, tells us how beautiful and bustling the town was before Katrina.

We rest briefly back at the headquarters and then I drive Hones to a fishing spot that Ryan told us about. It is off Route 23—*everything* here is off 23—over the canal where we saw an alligator earlier and up a dirt path to the hump of land on the Gulf side. I park on the dirt and we wander toward the tall grasses and reeds and, sure enough, there's the gap in the grass just where Ryan told us we'd find it. The sunlight is fading as we cut into the path and I'm a few feet ahead of Hones. I step over a log and am about to put my foot down when I notice it. Big and black, yellow underneath, semicoiled. It looks as thick as a python, and I'm pretty sure it's the biggest snake I've ever seen, including the boa constrictor Hones and I once saw in the Belizean rainforest. My foot freezes in the air, and the snake pauses, as if casually deciding whether to kill me or not, before slithering off into the brush. I jump back jittery.

"What the hell happened?" Hones asks.

I tell him what it looked like and we are pretty sure it was a cottonmouth, aka a water moccasin. I have seen plenty of snakes but for some reason this one leaves me feeling shaky. For all I love nature, I'm also fond of life.

"I've never seen you like this," says Hones.

My normal bluster has deserted me. I am blusterless.

"That thing was big," I say. "And scary."

When we head down the path again we walk more carefully. Hones finds a good spot and sets up a little camp for himself and I promise to be back to pick him up in a couple of hours. When I walk out I take extra care stepping over the log.

⌇⌇⌇⌇⌇

After my encounter with the snake, I drop by the lodge to visit Ryan. He asks where Hones is.

"At the spot," I say.

Ryan laughs out loud.

"The spot!"

He finds the name ridiculous, even though he was the one who told Hones about it. For Ryan any sort of land fishing here is junk fishing, not comparable to what you can reach by boat. But for Hones it works just fine. When I left him he was overjoyed with his new cottonmouth fishing spot and over the next day and a half he will catch a variety of fish both there and down in Venice—catfish, trout, redfish, drum.

Since Hones lost his job, he's been heading out to fish at Wachusett Reservoir every day, and he was reluctant to come on this trip lest he miss some of the last days of fishing season. The spot is a fine substitute: his own private Wachusett.

For my part, I'm glad to get some time alone with Ryan. Who knows when I'll be back? We sit at the long dining room table and tease each other about the coming election. That we can talk about it at all makes me feel better about the country.

"When we can only think in opposites, conservative and liberal, black and white, we can never reach any creative solutions," he says. "We have the same goals in mind, Professor, just a different route."

He tells me more about his childhood. He rues his lack of education. Straight to work after high school. Doing time at the chemical plant for twenty-one years. But still getting down here to fish and guide. "I was always driven, for some reason," he says. "Guiding was my part-time job. Only it was a full-time part-time job.

"I taught myself Spanish and I read a lot. And I've learned about nature and what makes her tick. And she's taught me. I know more about nature than most scientists, though they know the inside workings better than me."

Ryan, I understand, is not perfect. "Don't bullshit a bull-shitter," my father used to say. Ryan is putting on a show for me, the scribe. He is a big man with a big personality and a big story to tell and he is not going to apologize for that. I have dined with him and a group of fisherman while he entertained them with fish stories—violent, aggressive, sometimes bawdy stories that always starred Ryan Lambert—and remember that he mentioned offending someone and then added: "What are they going to do? They can't kick my ass and they can't get me thrown in jail for talking." I can see that the conversational mode he is most comfortable in is the monologue and I am fine with that.

But there are other parts, too, parts that I think I understand. Like him, I don't have much use for sleep. Like him, I had a father who died young. Like him, I am driven.

Maybe what I admire most about Ryan is the way he rebuilt after Katrina. It is a capacity I know I have in myself. I hope I don't say that boastfully. It's just that I have come to learn that the way I respond to devastation—personal or professional—is with a redoubling of effort and an angry determination to begin again. "I was possessed," Ryan said to me about rebuilding his lodge after Katrina. Exactly.

When Orrin Pilkey first told me that he thought the rebuilding of beach houses after storms was "societal madness," I nodded my head in agreement. I was with him in spirit. I am still with him on most things, but I can no longer nod about that one. If it is madness, it also neighbors one of our best human qualities. A determination, a fury, a fuel to make something, not just out of oneself, but out of the world. It is how things really get done on this planet.

We are sitting on the living room couch when I tell Ryan how impressed I am with his speechifying. When I listened to the tapes I'd recorded during the summer what I noticed most, other than his passion for this place, was the fact that there were almost no "uhs" or "ums" during the course of the interviews.

He seems pleased by this. He admits that he has considered running for office on a "Free the Mississippi" platform. I jokingly suggest that we should run for national office together, uniting the Red and Blue, and he agrees, laughing. But then I think about Ryan adding the burden of public office to the burden of all he does already, and I ask seriously: "Do you have the energy for that?"

As soon as it escapes my lips I know it is a stupid question. Here is a man who worked nights in a chemical plant and days as a bayou guide. Here is a man who has come back from utter devastation, his dreams shattered and flooded.

Here is a man who rebuilt those dreams, and his lodge, after Katrina, along with rebuilding a significant portion of Plaquemines Parish.

"Do I have the energy?" he repeats.

He looks at me dumbfounded.

"I *am* energy," he says.

On the last morning of my gannet trip to Nova Scotia I got up before dawn. I wanted to witness a spectacle of raw energy called the tidal bore. Within the course of half an hour all the waters near the Bay of Fundy begin a mad rush down dozens of tidal creeks, coming in powerful waves, and those waters fill the empty bay like a giant bathtub. It was summer, when the gravitational pull of the moon, and the sun, is strongest, making the tides all the more dramatic. In the middle of the night—at 3:30—I joined a ghostly procession of a dozen other guests streaming from their hotel rooms toward the water. We stared out at the empty mud chasm, about a hundred yards wide, where the Salmon River ran when it was in the mood. As if on cue came a noise like a train rumbling, the rumbling quickly growing louder. Then a wave, breaking in slow motion, charged toward us, filling the dry basin. While the wave wasn't huge—nothing you could surf, though I'd heard people try—the phenomenon, like that of a flash flood, was full of power and suddenness. Within ten minutes a river ran at our feet where a river hadn't been, gulls hitching a ride on the current while swallows darted and zipped, following the insects that followed the water. Once the basin filled, my fellow onlookers left, and I stayed alone as the river spilled over the brim. Then, in an exciting if not climactic finale, a kind of denouement, the river started to flow and burble in the opposite direction, the water reversing itself against the incoming tide. There was no one moment when you can say the river was really still. No moment either when you could clearly see that a switch had been thrown. But suddenly, on some parts of the water eddies formed and on other parts where sticks had been floating one way, they now floated another.

I felt electrified, and not just by the thrill of gushing water. Think of it. Approximately 100 billion tons of water flow in and out of the Bay of Fundy with every tide, the equivalent, scientists say, of the combined flow of all the freshwater rivers in the world (though one wonders how this could possibly be measured). As it turns out, the place has understandably inspired its own Jim Gordons and tidal energy plants are in the works that could produce 100,000 megawatts of energy, enough to power 100,000 homes.[55]

Wind, tides, sun. We need to use it all. And while we're at it there is another source we had better not forget about. As we search high and low for different types of energy, we had better not forget the human kind. How do we use that human raw material? This is a question every bit as pressing as "Will ethanol work?" How do we free up the sort of energy that Ryan has and let it work for, not against, the world? How do we tap it? How do we create *driven* environmentalists?

I'm not sure. But it doesn't seem to me we can do it softly at this point. It seems to me that the reason so many people admire Teddy Roosevelt, the reason I admire Ryan Lambert, is that they fight fire with fire. They are not anti-energy. They do not think they can force a river to run in straight lines. But they can nudge it, encourage it to move back into older paths. Human beings naturally want to be successful—and the urge to be successful provides the fuel of self-interest—but we must begin by redefining our meanings. Success can't just mean the biggest and most and straightest. Why couldn't you be the most ambitious man in the world and have your ambition be to see how much you can make of little?

It is vital to unite *excitement* with conservation. If we perceive "conserving" as dour, we will all turn the other way. We don't need saints and we don't need hushed quiet. We need Jim Gordon saying "Yes, I want to make money," and we need Jeff proud of piling oysters behind barrier islands, and we need Ryan talking excitedly about freeing the Mississippi. We need not just a hundred miles of oyster reefs, but a thousand. We need Orrin Pilkey taking glee in telling the truth about our barrier islands, despite the wrath of the locals. We need someone with sense, and a sense of nature, to take over the Corps of Engineers and show them, excitedly and once and for all, that straight isn't the only way to draw lines. We need young people excited about making something other than, or along with, money, who still believe in worlds other than the virtual.

~~~~

The next evening, our last, I drop Hones at his spot for a final night of fishing and head over to my own spot on the Mississippi. I park at the little boat launch on the river, the same one I visited this summer. The sun is dying back by Hones, over the millions of acres of wetlands, and that death is reflected in the sky over the river in pinks and blues. An osprey flies along the river's banks, riding the north wind that is blowing down clear from Minnesota and Wisconsin. The same wind brings the great final sweep of late migration, including loons, from the north.

I stare out at the river and I am happy, stupidly happy. We

are on the brink of one of the most divisive and acrimonious elections in our history—so what's new?—but for the moment that particular peep show, that carnival of charlatans, doesn't matter. Neither does the fact that oil is still washing up on shore or that the gills of the shrimp here are black as tar. No, what matters right now is the roseate spoonbill that flies overhead, its feathers almost exactly the same pink as the gaudy line of light below the blue clouds.

This show is here every night if you want to come see it. Wondrous things are always happening. I think of the people who first migrated to this land, whether by ice bridge or ship, and what they found: a land of wildly extravagant resources and wildly extravagant beauty. What we've done to that land and those resources would almost be comic if not for the tragedy, and we've done it with the alacrity of some crazy, sped-up film. Can we possibly redeem ourselves? *Fat chance,* says my cynical side. *Please, oh please,* says the other. By now I know that when Ryan talks about loss he is not just addressing the loss of species and wetland habitat. He is talking about the loss of a way of life. Not just hunting and fishing either, but the loss of a life connected in any way to the natural world. And as fewer and fewer of us are connected, fewer of us understand what we are losing. It's not about environmentalism. It's about this beautiful gray river and the osprey and the line of clouds and the roseate spoonbill. It's about what is best even if we sometimes forget what best really is. It's about wildness, a wildness that is still there inside our human chests and that vibrates like a tuning fork when we see a match for our wildness in the world.

The last streaks of sun blaze a gaudy pink. The water takes on a blue-black metallic sheen. Wind whips the sedge grass. Do we really want to destroy this? Do we really want to cut off the connection to the animals we are?

Maybe the trick, if you can call such a profound thing a trick, will be in redefining what we mean by self-interest. Environmentalists like to argue that saving an ecosystem is "useful," and, given their opponents' attitudes, this is perhaps a sensible approach. But it's such a limited way to think. By agreeing that that's the table the game will be played on, we tilt the game itself. Rather than *use,* it is the sheer wild uselessness of nature, the sheer nonutilitarian, unrelated-to-human uselessness, that is cause for celebration. I have always thought that nature was the source of my creativity, and the source of creativity for most artists, even those who never set foot on a beach or in the woods. But my thinking is evolving and I am moving beyond those inchoate ideas. I am coming to believe that nature *is* creativity. Not just a wellspring for humans but the thing itself.

Isn't it in our self-interest to hang on to that wellspring? Or at least to make sure that a path back to it exists? How odd that we are destroying the very thing that lets us imagine more. And by extinguishing things at the rate we are—there's no need here to drag out the well-known roll call of extinction and habitat destruction—we are doing no less than putting an end to creation. This may sound like overstatement. It isn't. For millions of years different species have evolved from each other in a thousand miraculous ways—*shrews becoming dolphins!*—and now we have said, "Enough is enough. Time for this messiness to end." We are busy neatening up the world, getting it organized, making it useful for humans, building our castles.

At my most pessimistic, I see no way of halting this. After all, we can never stop banging on things, can never stop making and improving. Most of us are no more capable of keeping still than beavers, whose own teeth impale them unless they constantly gnaw. So how can we, busy beavers that we are, have any hope of not chewing down all our trees and damming the world?

We are told that the castle builders and straight-line thinkers are the ones who understand the "real world." We'll see. The proof is in the pudding and if this summer is any indication, that pudding will be served up soon. We are also told that there is no time to get sentimental about nature, and anyway nature is in the way—it simply must go. And the scary thing is that most of us nod numbly and agree, if not in theory then in practice. But with every step forward we lose the path back. The more the straight-line thinkers win out, the less of a source is left for creativity. The less creativity, the less chance of getting back. When we kill the woods or beach we are killing possibilities. Our options, biologically as well as artistically, become limited. After all, you can't simply re-create dolphin or pelican or kangaroo.

I could go on but I will stop my preaching now. I am tired, weary. One of the things that straight-line thinkers like to do is to segregate, keeping everyone and everything in their separate cells. In this way, we can focus on the narcotics of our specialties: macramé or biochemistry or golf. In my field this means keeping art separate from politics, which is one of the rules of literature in the past century. It is a rule that I would, quite honestly, like to follow and one that I did follow for the first twenty years of my career. But it just doesn't seem possible anymore, what with the world ending and all. I'd like nothing more than to hole up in my garret and make art. But I can't. I am leaving this place tomorrow. But I will remain part of this mess forever.

ADAPTATION

On our last morning Hones and I wander across the street to the harbor, where the shrimp boats crowd the docks. The majority of the fleet has not budged during our stay. While the shrimping grounds are now open, bad weather and falling prices have kept the boats in the harbor. I walk out on the dock to a sound track of gulls, stopping when I get to the one boat with a running motor. The boat is named *Phong Fu,* and I strike up a conversation with the captain, a man named Tont, who, like many of these shrimpers, comes from Vietnam. He moved over in 1986 and his boat was named after his boy, Phong, who is twelve. I mention I have a girl who is seven and he tells me about his daughter, Kayla, who is six and was born after he got the boat.

"You'll have to get another boat to name after her," I say.

He laughs. "Maybe not. It's been a hard summer. We were closed down from April to September."

I don't ask if he has been compensated by BP. I say goodbye and walk back down the dock, gulls taking off before me, as Tont heads out to shrimp for the next six days.

Back at the headquarters I call up another local shrimp boat captain. Her name is Darla Rooks and she has been outspoken during the crisis. It turns out she is speaking to me by cell phone from her boat. Like Tont, she is heading out on the water today, though she's just going for a couple of days, hoping to get back in time for trick-or-treating.

"I worry that some people are shrimping in illegal waters, too close to the rig. They do that and one person eats a tainted shrimp somewhere in Indiana and we are ruined. It could shut us down.

"Some people claim I shouldn't be saying these things, but in private people will tell you they opened up the shrimping too early. They should have kept it closed until they could be sure it was safe. At least shrimp are better than oysters. They sank dispersants right on top of my oysters. They wiped them out. They say 'Everything's good, everything's good,' but we know everything is not good. They just sent me a check to the wrong name in the wrong amount. But that's not the real problem. The real problem is that I want to live here because this is my home, but they ruined this place. I spent my whole life learning this place and I used to say I would never move anywhere else. If I move I gotta learn a whole new place. But now I think I don't have a choice. I think I should move like other people are doing."

I might as well be back in Nova Scotia with Keith. Will this place soon be a ghost town, like Sydney Mines? Will it be hollowed out? Will the people who live here drift on to new places, new lives?

⁓

For Darla Rooks the ledger sheet of loss and gain is crystal clear. She has been asked to give up almost everything: tradition, profession, home.

For the rest of us it is more complicated. During the fall, some other professors replied to my inquiry about ways to connect the oiled pelican to the world. They didn't quite manage to link the birds to Saturn, though one English professor described seeing pelicans by the launch site of Cape

Canaveral. A philosophy professor recalled that the pelican was a religious symbol and sent along this Wikipedia entry: "In medieval Europe, the pelican was thought to be particularly attentive to her young, to the point of providing her own blood when no other food was available. As a result, the pelican became a symbol of the Passion of Jesus and of the Eucharist." And, along the same lines, another professor offered up Psalm 102, which ends: "I am like a pelican of the wilderness."[56]

All of it went in my file, though I made special note of the idea of the birds feeding their blood to their young. As natural history it's hogwash, but symbolically I can see the pelican as the offering we have sacrificed at the altar of oil, down in this body of water that is our national sacrifice zone.

There's another way to look at it, though. Maybe the true offering has to come from us, in response to what has occurred. At the very least, the idea of sacrifice, which seems so outdated and quaint, has to be revived. In an age of instant gratification, why ever give anything up? Perhaps because by giving up we gain something greater. Sacrifice is a time-tested concept, of course, the opposite of radical, but one that has fallen almost completely out of fashion.

The "something greater," in this case, is the world itself. I still love that world, and I don't think I am alone in this. I want bird migrations to remain part of my life. Since I celebrate messiness, I am certainly not suggesting that we all start behaving perfectly and create a litterless and pristine world. We need energy, we need food, we need fun. All I'm suggesting is that when we make the countless decisions, big and small, that we make every day, we remember that there are priorities other than efficiency, comfort, and profit. If, on the other hand, we make those cold and soft values of ease our gods, those gods will exact a price. We

may gain some ease and comfort, but we lose their opposites, and those opposite qualities are part of what is best in
the world. I am thinking about sacrifice—real sacrifice—for
a greater purpose, and I am thinking about effort, and I am
thinking about joy. I am thinking about the art of getting
and making food, as well as the art of getting from place
to place, of calculating our energy consciously just as a teal
or gannet does instinctively. I am thinking about how we
would act if we truly spent some of our time imagining
what life will be like for our children and their children and
not acting as if we are living at the end of time.

It's true we may never match oil, miraculous oil, for sheer
ease. But since when has ease been everything? As any adult
knows, we sacrifice for our children. Is it too much to think
we might act like adults? This means not simply dumping
off our burdens on others. What could be more childlike
than to pretend that our actions don't have consequences?
Rather than self-sacrifice, we have decided that it is okay
to simply sacrifice places where we aren't, that it's okay to
give up parts of the world and peoples, the Gulf, Alaska, and
of course, our specialty, other countries. It is okay as long
as we don't have to see these places, or think about them
too much (which is one reason the media has skulked off).
And what do we gain by this great sacrifice? We get to live
exactly as we have been living. Which of course is a great
good. We must do what we have done and we must keep
doing it because . . .

Well, because why? Is this lifestyle of cars and computers
and waste so wonderful? Does it fill us with deep satisfaction? It must if we are willing to sacrifice miraculous places
and time-tested ways of life.

Think instead if we sacrificed a little. Think instead if we
tore off our blinders and stopped pretending that "down

here" has nothing to do with where we ourselves are. Think if we looked directly at the thing and admitted that we were complicit and started from there. Think if we stopped pretending that this is a reality show and we are simply uninvolved viewers watching it from the couch. Think if we took our considerable energy and used it, channeled it, toward something both old and new.

<center>~~~~~</center>

We drive north to New Orleans, stopping only once when Hones spots some white birds soaring over the water. I am hoping for gannets, but the consolation prize isn't bad: more white pelicans, seventeen or so in a tight formation, bank and curve over the Gulf waters. They flip, like a card trick, from bright white to black as they turn.

"I could get used to seeing them," I say to Hones, though we both know I won't, since the birds don't breed or migrate in the East.

New Orleans is packed, Bourbon Street crowded not just with the usual throng of humanity but with Steelers' fans here for tomorrow's game against the Saints. It is likely that they are seeing things they never saw back in Pittsburgh, like the Saints fan who wears the shirt of Ben Roethlisberger, the Steelers quarterback who has been accused of sexual assault. A naked blowup doll is attached below the fan's waist so that the doll seems to be in the act of performing fellatio. Ah, Bourbon Street.

Hones and I wander over to French 75, but when we find it doesn't open until six in the evening, we move over to the casual dining section of the same establishment. We eat boudins (pronounced, I learn from mispronouncing, *Boo-dan*), delicious meat pies, and Hones concludes his scientific Gulf seafood sampling with a dozen raw oysters. Before we leave the city we wander down to the river. The sun is blazing and people are out walking and a riverboat named *Natchez* blasts its horn while a saxophonist plays "Somewhere over the Rainbow."

We drive to Biloxi and stay across the street from the casino. Hones watches as I make three hundred dollars and then lose four hundred dollars playing craps. The next morning we search for gannets on the boardwalk under the highway, and I think I see one though it may be wishful thinking. We wander down the docks and look at the boats of the mostly Vietnamese shrimping fleet of Biloxi, serenaded by "Love the One You're With." This is because the fleet, looking ancient and still proud, now rests in the shadows of the giant outdoor speakers of the Hard Rock Casino.

It is time to get back to Pensacola, to the airport, but I talk Hones into one more side trip. I am on a wild gannet chase and I can think of no better place to look for them, and to end the trip, than where it all started at Fort Pickens, aka Tarball Beach. Sure enough the workers are still there, though I must say I have more sympathy with them after talking to Sergio. We head down to the beach with our binoculars and the trip's final two beers. It is beautiful and calm and the water tropical green, which means it is not good gannet weather. It occurs to me that they may not even be here yet, due to their long offshore migration after leaving late from their rock stacks near Newfoundland.

I am willing to accept my own theory—we missed them—and leave it at that, but as we are walking off the beach I bump into a ranger who is heading down to the water.

"Do you know birds?" I ask.

"I know they fly," she says.

"Do you happen to know if you get northern gannets here?"

"Yup, we get them."

"So maybe they just aren't here yet?"

"No, they're here. I know because they just took an injured one in at Perdido Bay. First one of the season."

So they are here.

We walk on, back to the car, and drive to the airport.

I forgot to ask if the bird was oiled.

⸻

Home again. Before heading back to Boston, Hones stays for a day, which thrills Hadley. She rejoices in his resemblance to Santa Claus.

Life pours back in and sheer busyness takes over, the Gulf fading. But over the next few weeks I make a point of getting out to the beach to look for gannets. I see none during my first few expeditions. Meanwhile back at home I eye the spot where I will build my cabin by the marsh, where I will do my own personal math, not pretending that that math will somehow save the world. Deep down I still prefer Thoreau the wild man to Thoreau the economist and I will never be as strict as Henry. But that doesn't matter; that isn't exactly what I am after. What I am after, I'm just beginning to understand, is knocking over the statue called comfort, and seeing if sacrifice is still a possible virtue. But that isn't it exactly either. Less simply, I am coming to see that I am just as connected to this world as the brackish waters in my creek are, and that some sacrifice must be made, either by me or someone or something else, for my comfort.

Right before Thanksgiving I head to the beach again. The wind has kicked up and the cold come in and I finally see what I've been waiting for. It is a day of churning surf and all the birds—pelicans, gulls, gannets—are plunging. But if it is a show of excess then there is no question which bird is the most excessive. I'd always thought of gannets as a cold-weather bird, a wind bird, and they seem to practically exult in the wild weather, turning and dropping like fighter pilots, as if showing off. To watch them dive is to watch a great festival of excess: the birds plunge by the hundreds for fish—*thwuck, thwuck, thwuck*—as if they believed in nothing more than wild abandon.

Gannets fly in the face of Thoreau's less-is-more thinking. They long ago embraced a species-wide philosophy of excess. Theirs is the math of more, and they repeat their high dives again and again. Because they dive so much they need more fish and because they need more fish they dive so much. And it works. There are plenty of fish and they have plenty of energy thanks to the plenty of fish. What might look like squandering is in their case strategy.

You have probably made the mental jump before me. Americans have long been proud gannets. And why not? Rather than beat ourselves up about this fact, why not admit that for many years the math made sense. It worked for us. We stumbled upon a wide-open, relatively sparsely populated country, a country full of trees, animals, fossil fuels, gold, you name it. How were we expected to respond given the circumstances? With caution and frugality? My field guide calls gannets "gluttons" and they have to be to supply their nonstop internal engines. It's a crazy way to live, though it seems to work for them, and should continue to work for them as long as there are fish in abundance.

I am not wagging my finger here. For my part, I've always been a squanderer, charging ahead, pushed by my own ambition, rarely pausing. For one thing, it seems more *exciting*:

who wants to go through his one life bored? And if I have lived a gannet life and we have long been a nation of gannets, what of it? As long as there are fish aplenty, why change?

And here's another question: Is it really possible for me, and for the rest of us, to be happy with less? Is it possible to make sacrifice as attractive a virtue as ease? I'm not sure. You could argue that *Homo sapiens* are about as likely to change our ways as gannets: we are what we are. But even the dimmest of us seem to have become aware of certain connections—between our consumption and the world— that almost no one considered fifity years ago. Now, with our luck running out along with our resources, we are per-haps starting to notice things we didn't notice then—didn't notice or pretended not to notice. Which in turn creates a cognitive dissonance between the way we live and the way we know we *should* live. It would be easy enough to shrug and say, "Hey, we're gannets, what can we do?" Except for the fact of that singular human trait that some, like the evo-lutionist Stephen Jay Gould, claim defines us: our adaptabil-ity. The fact that we can change over a lifetime, not just over evolutionary eons. I don't want to get hokey here and say that something is being asked of us. But maybe something is being asked of us. And maybe this time around we'll dig-nify what is being asked with an answer. Even if the answer is: "Well, what the fuck do we do now?"

I don't have any answers, but I do know the question is out there. If my waiter at Applebee's could preach about energy and connections to me, you know it's in the air. It has taken a long time for most of us to understand that our wild national orgy might just be over. Right now thousands of gannets are returning to the Gulf, where they will plunge deep, again and again. What will they find? Will there be enough fish and, if there are, will those fish carry secret contaminants that won't

reveal themselves for years? Stay tuned. But amid all the uncertainty one thing is certain: if the seas are empty or the fish
subtly poisoned, the birds will not have the luxury of our
species, that of fast adaptation. Gannets, unlike us, have no
other way of being.

The Testimony of Captain Ryan Lambert
Director Louisiana Charter Boat Association
Before the
Subcommittee on Insular Affairs, Oceans and Wildlife
of the House Natural Resources Committee
Hearing on Our Natural Resources at Risk
"The short and long term impacts of the *Deepwater
Horizon* oil spill."
June 10, 2010   10:00 am

Dear –

I am deeply grateful to the Subcommittee for the opportunity to testify at this hearing, and to explain the impact that the *Deepwater Horizon* oil spill is having on my community, my business, and my way of life.

I am a member of Ducks Unlimited and the Coastal Conservation Association, and sit on the Board of Directors for the Louisiana Charter Boat Association, as well as being president of Cajun Fishing Adventures. As a professional fishing and hunting guide with twenty-nine years of experience, I have built one of the most successful fishing lodges in the state of Louisiana. I am licensed by the Coast Guard, and I have been hunting, fishing, trapping, and shrimping in South Louisiana all of my life.

As the years have passed, our way of life has been increasingly threatened due to the erosion of our wetlands. These wetlands are a place where our unique culture has existed for generations. They are our home, and we value working in the rich Gulf waters.

The people of Louisiana have been stereotyped as being "backwards" or "behind times." The truth is, we are just salt-of-the-earth Americans. Americans who are not afraid to roll up our sleeves and make a living off of the land. When our shrimp season closes or our crabs aren't giving, we adjust to find another way to make our money off the land. We don't run to the unemployment line and we don't seek the help of agencies. Unfortunately, now that our shrimp boats are in dock and our crab traps are on the bank, there aren't any sportsmen wanting to come down to fish or hunt waterfowl with guys like me.

With millions of gallons of oil entering this fragile eco-system from the oil spill of the *Deepwater Horizon,* never before has our national treasure been in more jeopardy than it is now. It is apparent that it is time for us to turn to you for the help we need to save our precious wetlands and our way of life.

For far too long, Louisiana's restoration projects have been held back due to red tape and political bureaucracies. It is time for someone to step to the plate and reconnect the Mississippi River to the marshes it sustains. This disconnect is at the root of our problem. A spotted owl can stop the logging industry. An endangered mouse can halt a housing development. But we lose the size of a football field every thirty minutes as we sit back and let the greatest estuary in North America go by the wayside. This estuary supports the vast majority of south Louisiana with its great abundance of resources such as oil, seafood, fishing, and hunting. It is an economic engine in itself.

The Mississippi River is one of the most highly engi-neered in the entire world and provides great benefit to the nation's economy at Louisiana's expense. For years the Corps of Engineers has dredged the river and put the sediment in hopper barges taking it offshore to dump it in the Gulf in-stead of putting it to good use in our wetlands. The reason given was that it is not cost effective to use it in the marsh. How many millions of dollars do we need now because we did not spend the extra money to use this resource wisely?

Twenty-five years ago, the restoration of a major portion of the Plaquemines Parish shoreline, the Shell Island project, was estimated to cost thirty-five million dollars. Unfortunately, it was not completed at that time. That same project has again been under study by the Corps of Engineers for over five years. The current estimated cost is $250 million. The time for studies has past.

The Corps is primarily a flood control and navigation agency, and has no mission or procedure to elevate the restoration of south Louisiana to levels of equal importance as its traditional missions. The precedent is no action. We need a new precedent. We need to take extraordinary action, which will involve risk and uncertainty. We need to send the Corps a new mission. A mission that is at least equal to the navigation of the lower Mississippi, a mission of restoration!

As we did after Katrina, we are again watching our military helicopters flying sandbags trying to plug the large gaps in our coastlines. Had we taken control of our river and sediment years ago, we would not have to protect ourselves from the large plumes of oil lurking off the coast. We would not be in the fix we are now in. This is the fourth time in recent years we have felt the sting of our failures. Without taking into account the hurricanes and this oil spill, we are losing countless acres of wetlands every day. The time has come to save our national treasure.

Other states refuse to drill off their coast, yet they allow Louisiana to take the hit when something like this oil spill happens. Louisiana has been refused royalties due to the state for drilling in our fragile ecosystem. Now after all the years we have been supplying the country with 30 percent of domestic oil from the Gulf, we will start getting well-deserved royalties in 2016. This is too little, too late. This money should be sent to Louisiana immediately. The money should be sent to fund programs such as the Coastal Wetlands Planning, Protection and Restoration Act (CWPPRA) and used solely for the restoration of our abused coast.

We don't know for sure the long-term effects that the dispersants and the millions of gallons of oil are going to have on our marshes. We do fear that after the visible oil is cleared and the news media is gone, we will be left to wait for Mother Nature to heal herself. We will be left without a way to make a living and our wetlands will just wash away.

It seems that many people refuse to see the big picture of what is really happening. While the loss of pelicans and turtles are devastating scenes, the real damage is going on inside the marshes. These marshes serve as the nursery to 20 percent of the nation's commercial seafood. The eggs and larva of shrimp and crabs, the spat from oysters, as well as the young of many of our fish species are being killed by the millions. Without these young and the invertebrates that they feed on, Louisiana and our way of life will be changed forever. All life starts at the bottom of the food chain, this is where the most damage will occur when the oil and dispersants cover our waters.

Also, when one of the greatest natural spectacles in North America starts in late August, with the migration of our waterfowl and other wetland birds, if the oil is not cleaned up by then, this alone will be truly a national disaster of epic proportions. This migration will send some fifteen million waterfowl passing through south Louisiana. A great percentage of them will winter in Louisiana until the spring winds call them back to the north to nest. The wetlands of the Gulf Coast comprise the most important wintering area for waterfowl and many other wetland-dependent migratory birds in North America. Perhaps 50 percent of the ducks in the north migrate through or winter in Gulf Coast wetlands. The spill will devastate these birds, some of which are already threatened. Everyone has seen the photos of pelicans and other shorebirds covered in oil. Imagine photos of millions of waterfowl and other beautiful birds, covered in black. My other fear is that the small animals and invertebrates as well as many aquatic grasses will not be present. These are the fuel sources that take many of these birds to Central America to winter. Plaquemines Parish, where I make a living, contains 14 percent of America's wetlands. A major percentage of the Mississippi flyway waterfowl winters here. This is ground zero for the *Deepwater Horizon* oil spill. If we lose 50 percent of these waterfowl, the economic impact will be felt from Alaska and Canada and throughout the central United States for many years.

I sit here preparing my written testimony, having just returned from a visit to one of our completed restoration projects. I think about how optimistic I was this morning before arriving on the beach. I thought maybe BP was right—that it is not coming inshore because after forty-seven days, I hadn't really seen the giant oil slicks everywhere. Now I have lost the wind from my sails after seeing millions of tarballs rolling in the surf. Not only was every tarball covered with small dead clams, but just under the surf are millions of these clams covering the beach. This is just the start of the death that we will be seeing in the future.

By BP putting the dispersants on the oil, it has sunk out of sight of the cameras. The oil is there, millions of gallons of it. It is just starting to make its way to the Louisiana shores. My walk took place ninety miles from the *Deepwater Horizon*. Areas closer to the oil well don't have a beach to protect it from the oil balls coming into the marsh. They are underwater where you can't see them, but they are there. This is just the start of what's to come. The oil will be coming from the depths for years, not floating on the surface but out of sight. It is not too late to rebuild our coast. We need to open up the Mississippi to the marshes and let it do its job the way nature intended it to. There is a happy medium between navigation and restoration. We need to find that place and find it fast. Now is not the time for more studies. It's time to get the river flowing through the natural channels that still exist. Sure there will be shoaling in places, but it doesn't take the whole river to navigate to New Orleans.

I thank the Subcommittee for letting me share my thoughts on our great Mississippi Delta. I would also like to take this opportunity to invite each and every one of you to come down and let me show you in person just what I am talking about.

Thanks,
Captain Ryan Lambert

NOTES

1. John Muir, *My First Summer in the Sierra* (New York: Houghton Mifflin Company, 1911), 211.

2. Peter Lehner, *Deepwater Horizon: The Oil Disaster, Its Aftermath and Our Future* (New York: OR Books, 2010), 62.

3. Scott Weidensaul, *Living on the Wind* (New York: Macmillan, 2000), 26.

4. William J. Broad, "Nuclear Option on Gulf Oil Spill? No Way, U.S. Says," *New York Times,* June 2, 2010.

5. Lehner, 67.

6. John M. Barry, *Rising Tide* (New York: Simon & Schuster, 1998), 10.

7. Justin Gillis, "Giant Plumes of Oil Forming Under the Gulf," *New York Times,* May 15, 2010.

8. "The Spill," first broadcast October 26, 2010 by PBS, *Frontline* (Boston, MA: WGBH/Boston). Produced by Martin Smith and Marcela Gavira.

9. Lehner, 69–72, 123.

10. Lehner, 126.

11. The World Factbook (CIA) https://www.cia.gov/library/publications/the-world-factbook/rankorder/2174rank.html (accessed June 5, 2011). The United States consumes 18.7 million barrels per day. Total oil spilled during the Deepwater Horizon event was 4.9 million barrels.

12. Jessica Resnick-Ault and Katarzyna Klimasinska, "Transocean Rig Sinks in Gulf of Mexico as Coast Guard

Looks for Survivors," Bloomberg L.P., April 22, 2010, http://www.bloomberg.com/apps/news?pid=newsarchive &sid=aHylLWhmGcI0 (accessed June 5, 2011).

13. Lehner, 69.

14. John M. Broder, "Obama to Open Offshore Areas to Oil Drilling for First Time," *New York Times,* March 31, 2010.

15. Rocky Kistner, "BP Oil Disaster Index (Vol. 1)," *Switchboard: Natural Resources Defense Council Staff Blog,* September 21, 2010, http://switchboard.nrdc.org/blogs/ rkistner/bp_oil_disaster_index_vol_1.html.

16. Christine Dell'Amore, "Coast Pipelines Face Damage as Gulf Oil Eats Marshes?" *National Geographic News,* May 25, 2010, http://news.nationalgeographic.com/news/ 2010/05/100525-gulf-oil-spill-pipelines-science-environ ment (accessed June 5, 2011).

17. Abby Goodnough, "Wind Farm off Cape Cod Clears Hurdle," *New York Times,* January 16, 2009.

18. Cape Wind, http://www.capewind.org/article31.htm (accessed June 5, 2011).

19. Sean Corcoran, "U.S. Approves First Offshore Wind Farm," *All Things Considered.* National Public Radio (Washington, DC: NPR, April 28, 2010).

20. Reg Saner, "Snow," *New England Review & Bread Loaf Quarterly* 11, no. 2, (Winter 1988): 129.

21. Shelley DuBois, "Update: BP's advertising budget during the spill neared $100 million," *CNN Money,* September 1, 2010, http://money.cnn.com/2010/09/01/news/ companies/BP_spill_advertising_costs.fortune/index.htm ?section=magazines_fortune&utm_source=twitterfeed &utm_medium=twitter&utm_campaign=Feed%3A+rss %2Fmagazines_fortune+%28Fortune+Magazine%29 (accessed June 28, 2011).

22. Richard Wray, "BP shares rise as Deepwater Horizon well repairs progress 'as planned,'" *Guardian,* July 12, 2010.

23. Ylan Q. Mui and David A. Fahrenthold, "Gulf Seafood Must Pass the Smell Test," *Washington Post,* July 13, 2010.

24. V. M. Janik, L. S. Sayigh, and R. S. Wells, "Signature Whistle Shape Conveys Identity Information to Bottlenose Dolphins," *Proceedings of the National Academy of Sciences of the United States of America* 103, no. 21, May 23, 2006: 8293–8297.

25. Joel K. Bourne, Jr., "The Spill," *National Geographic,* October 2010, 53.

26. "BP: 'Small People' Matter to Us, Chairman Carl-Henric Svanberg Says," *Huffington Post,* June 16, 2010, http://www.huffingtonpost.com/2010/06/16/bp-small-people-matter-to_n_614705.html (accessed June 5, 2011).

27. NOAA's State of the Coasts, http://stateofthecoast.noaa.gov/population/welcome.html (accessed June 5, 2011).

28. A. R. Ammons, "Dunes," in *Selected Poems* (New York: W.W. Norton, 1986), 51.

29. John Keats, *The Selected Letters of John Keats* (Cambridge: Harvard University Press, 2005), 60.

30. NOAA's State of the Coasts, http://stateofthecoast.noaa.gov/population/welcome.html (accessed June 5, 2011).

31. Sarah Parsons, "Deepwater Horizon Spill Could Kill America's Wild Shrimp Industry," *Change.org,* May 14, 2010, http://news.change.org/stories/deepwater-horizon-spill-could-kill-americas-wild-shrimp-industry (accessed June 28, 2011).

32. "Bird Impact Data and Consolidated Wildlife Reports," US Fish and Wildlife Service: FWS Deepwater Horizon Spill Response, http://www.fws.gov/home/dhoilspill/collectionreports.html (accessed June 28, 2011).

33. "The Intergovernmental Panel on Climate Change Fourth Assessment Report: Climate Change 2007," http://www.ipcc.ch/publications_and_data/publications_and_data_reports.shtml (accessed June 6, 2011).

34. The New York City Office of Emergency Management, "NYC Hazards: NYC Hurricane History," http://www.nyc.gov/html/oem/html/hazards/storms_hurricanehistory.shtml (accessed June 6, 2011).

35. United States Landfalling Hurricane Probability Project, http://landfalldisplay.geolabvirtualmaps.com/, (accessed June 6, 2011).

36. Scott A. Mandia, "The Long Island Express," http://www2.sunysuffolk.edu/mandias/38hurricane/hurricane_future.html (accessed June 6, 2011).

37. Jennifer Peltz, "Hurricane Barriers Floated to Keep Sea Out of NYC," Associated Press, May 31, 2009.

38. Klaus Jacob, "Time for a Tough Question: Why Rebuild?" Washington Post, September 6, 2005.

39. National Oceanic and Atmospheric Administration, "June, April to June, and Year-To-Date Global Temperatures are Warmest to Date," July 15, 2010, http://www.noaanews.noaa.gov/stories2010/20100715_globalstats.html (accessed June 6, 2011).

40. Eric W. Sanderson, Manahatta (New York: Abrams, 2009).

41. Metropolitan East Coast Assesment, "Climate Change and a Global City: An Assessment of the Metropolitan East Coast Region," http://metroeast_climate.ciesin.columbia.edu/ (accessed June 6, 2011).

42. Scott Weidensaul, quoted in Ornithology, ed. Frank B. Gill (New York: Macmillan, 2007), 273.

43. Barry D. Keim and Robert A. Muller, Hurricanes of the Gulf of Mexico (Baton Rouge: LSU Press, 2009).

44. Gary Snyder, "Four Changes" in *Turtle Island* (New York: New Directions, 1974), 101.

45. Conrad Aiken, *The Letters of Conrad Aiken and Malcolm Lowry, 1929–1954* (Toronto: ECW Press, 1994), 197.

46. John Keeble, *Out of the Channel* (Cheney: Eastern Washington University Press, 1999), 323.

47. Charles Wohlforth, *The Fate of Nature* (New York: Macmillan, 2010), 292.

48. Keeble, 325.

49. Keeble, 11.

50. Henry David Thoreau, *Walden,* ed. Jeffrey S. Cramer (New Haven: Yale University Press, 2004), "Economy."

51. Justin Gillis, "U.S. Finds Most Oil From Spill Poses Little Additional Risk," *New York Times,* August 4, 2010.

52. Richard Harris, "Scientists Find Thick Layer of Oil on Seafloor," *All Things Considered.* National Public Radio (Washington, DC: NPR September 10, 2010).

53. Bourne, Jr. "The Spill," 58.

54. Bradley Blackburn, "BP Cover Up? Plaquemines Parish President Billy Nungesser Says Oil Giant Lies About Oil Spill Cleanup," *ABC World News with Diane Sawyer,* August 2, 2010, http://abcnews.go.com/WN/bp-oil-spill-plaquemines-billy-nungesser-accuses-bp/story?id=11307871 (accessed June 6, 2011).

55. "Tidal Energy: Bay of Fundy," *Renewable Energy Development,* April 19, 2008, http://renewableenergydev.com/red/tidal-energy-bay-of-fundy-canada/ (accessed June 28, 2011).

56. "Psalms 102:6 (King James Version)" in *Biblos,* http://kingjbible.com/psalms/102.htm (accessed June 28, 2011).

ACKNOWLEDGMENTS

The list is long of folks to thank, but let's start, logically enough, at the beginning:

I was in northern Vermont at a conference, in the early days of the BP spill, and it was there that I met Bethany Kraft, Executive Director of the Alabama Coastal Foundation, without whom this book would have been impossible. When she was not in classes at the conference she was on the phone with BP officials, and over the next year she gave me guidance, advice, insight, and a bed in the guest room of her home in Mobile, Alabama. To her I offer my first thanks. At the same conference I also met Richard Waller and Jess Bloomer, who were already thinking and writing about the Gulf, and who helped me sharpen and clarify my own thinking.

While it's true I had already begun to ponder the spill by the time I got home to Wilmington, North Carolina, I had no plans to head down to the Gulf itself. That changed at a small party at a friend's house, where the writer John Jeremiah Sullivan urged me to do so. I would have suspected that this was an act of singular generosity, were it not for the fact that I had heard from other writers of how John had done similar things for them: editing, nudging, and encouraging, often more aware of what they should be working on than they themselves were. The next step was finding support, funding, and structure for the trip, and these were provided by *OnEarth* magazine and the Natural Resources Defense Council. News Editor Scott Dodd hired me to write a piece on bird migration and began publishing my blogs from the road. He also pushed and prodded to make the work better. George Black, my brilliant editor at *OnEarth,* later encouraged me to refine and expand my thinking about the Gulf. The NRDC—and especially Rocky Kistner, who ran the Gulf Resource Center—provided me with contacts, leads, and a place to stay while in Buras. For this I am deeply grateful. Another person who went out of his way to point me in the right direction was the filmmaker Randy Olson—many thanks to him.

Once I reached the Florida Panhandle I began, increasingly, to rely on the kindness of strangers. Pamela Thacker and the rest of the ranger staff at Fort Pickens were generous and encouraging, and in my first days, when I was questioning what right an outsider had to write about this topic, they assured me that I was doing okay. Also in Fort Pickens I bumped into James Emery, who talked freely of his personal response to the oiled waters and was, for a day at least, a friend. In Pensacola I also spoke with Frank Patti and Alice Guy, who were honest and direct about how the oil had impacted the seafood business.

This book began as an experiment in the potential of blogging, and so I can't forget that though I was driving solo through the Gulf, I was never alone. I lived full days but most mornings I was up before dawn to report back to both *Into the Gulf,* my NRDC blog, and *Bill and Dave's Cocktail Hour,* the website I started with Bill Roorbach, a great friend and writer. The blog group became a kind of society that traveled with me, and often the comments I got back, including those from the writer and scientist Eva Saulitis, who had experienced the Exxon Valdez spill, became part of the adventure.

In southern Louisiana, I met Captain Sal Gagliano, who gave me a tour of the oiled marshes. I also met the filmmakers from Jean-Michel Cousteau's Ocean Futures Society. I can't say enough about their low-key professionalism, good humor, kindness, and generosity far above and beyond the call of duty. I am particularly indebted to Holly Lohuis and Brian Hall, and I thank Nathan Dembeck for captaining the wheel of the Zodiac and trying to fix my car. Of course the reason I met the Cousteauians was that I was staying at Cajun Fishing Adventures Lodge, where I had the pleasure of meeting Lupida Sauseda, who gave me food, drink, and laughter, and Ryan Lambert, without whom there would be no story. As I think I make clear in the text itself, Ryan was instrumental in helping me think about the eroding wetlands and the culture, and it was through his love of the land that I got my first glimpses into the place itself. My debt to him is deep.

I also need to thank Anthony and his uncle, Lyn, for letting me spend the night at their fish camp. For my chapter on dolphins I need to thank Ann Pabst at the University of North Carolina Wilmington, and Robert Smith, who guided me along the waterway from North Carolina to South Carolina. My other companion on that trip was Rory Laverty, who, more and more, has become my co-adventurer, always up for a paddle and a beer. Also, for earlier adventures on the water, I'd like to thank Douglas Cutting, whose love for these shores is contagious. Both Doug and Rory are former students and it is remarkable how much I rely on this group for support and friendship. Particularly invaluable has been the work of Doug Diesenhaus, who helped with the research and whose calm genius provides a wonderful counterbalance to my more erratic nature. Another student, Arianne Beros was a huge help when it came to the editing.

Through Bethany Kraft, I got to know Bill Finch, who gave generously of his time and took us on an educational tour of both Grand Bay and Dauphin Island. I also need to thank Jeff DeQuatrro of the Nature Conservancy, who took us out by boat to see the western end of Dauphin Island, as well as Cat Island, to see white pelicans. In Bayou La Batre, I came across Jim Duffy, who gave an elucidating lecture from the bow of his small boat, and Miranda, my wise-beyond-her-years waitress at the Blue Heron Café. Thanks to her for distilled nuggets of wisdom and, of course, for the gravy.

For my trip to New Orleans I need to thank Kristian Sonnier and Kevin McCaffrey, who gave me valuable insiders' insight into that magnificent and troubled city, as well as pointing out good places to go to dinner. Driving into the city I had no idea how to get into the French Quarter, so I called Kate Sidwell and she acted as navigator. This was metaphorically apt, as Kate and her husband, the irrepressible Steve Sidwell, have provided guidance, encouragement, and an occasional roof over my head over the last few years.

Throughout my time in New Orleans and southern Louisiana, I was thinking hard about coastal issues and sea level rise, so perhaps it is time to thank the great Orrin Pilkey. Orrin and I have now traveled together down the Outer Banks and through the streets of New York, and there was hardly a moment when I was not learning something new or, just as importantly, laughing. He is a model for anyone who aspires to be a scientific advocate: aware of the risks of and criticism inherent in that position, but with an understanding of just how important it is to fill the role in these dire times. I salute him.

Kerry Emanuel at MIT helped to educate me about the ways warming waters intensify storms. Sergio Pierluissi helped me understand the sheer strangeness of working for the clean up crews, and Scott Weidensaul, whose writing I have long admired, answered my questions about the stresses of migration. Also valuable were Paul Spitzer, Frank Moore of the University of Southern Mississippi, Joel Bourne of *National Geographic,* as well as Chris Wood, Ken Rosenberg, Brian Sullivan, and Laura Erickson of the Cornell Lab. On Dauphin Island I interviewed John Dindo and Ken Heck, and their insights and wisdom became part of this book. And of course, I need to once again thank Alan Poole, who remains my go-to birdman—always quick to answer my questions when I'm in a pinch.

I also owe a debt of gratitude to Larry Cahoon and the other members of the UNCW Evolution Learning Committee. And to the irrepressible Walter Brooks, who set up my lunch with Jim Gordon.

I thank Mark Honerkamp, for continuing to work as my Sancho Panza, despite the lousy pay. Over the years I have learned that people are deeply sensitive about being written about, and Hones, thankfully, is the one exception.

And I thank my mother, Barbara Gessner, who has supported me in every way possible. Lucky for me I have not just one, but three parents who support me: As the years go by I realize what pure fabulous luck it is to have supporters of my writing like Georges and Carol de Gramont.

I am also fortunate to be a member of another family, the creative writing department at University of North Carolina Wilmington, chaired by Philip Gerard and co-chaired by Megan Hubbard. They, too, have supported me as I've roamed. There are too many of you in the program to name, but you know who you are. Being part of this community has meant the world to me.

My thanks to Russell Galen, my agent, who read the early blogs and saw a spark in them, and who kept his enthusiasm, and cool, when it looked like the world had been flooded, not just with oil but with book proposals about oil.

And huge thanks to Patrick Thomas, of Milkweed Editions, who saw, early on, the potential for this to be a book about much more than the spill, and who stuck to that vision throughout a year of hard and nearly constant work. He pushed me to places I didn't want to go but that I am glad I went to, making the book more broad and expansive.

Finally I need to thank my wife, Nina de Gramont, whom this book is dedicated to and who supported my insane decision to get in the car and go. She has always been my greatest supporter, as well as my editor, inspiration, and friend. My gratitude to her is inexpressible.

DAVID GESSNER is the award-winning author of eight books and countless essays about the wild world. Winner of a John Burroughs Award and selected for publication in *The Best American Nonrequired Reading,* his uniquely rambunctious style has been redefining what it means to write about nature for the past twenty years. He teaches creative writing at the University of North Carolina Wilmington, where he founded the award-winning journal, *Ecotone.*

More Nonfiction from Milkweed Editions

To order books or for more information,
contact Milkweed at (800) 520-6455
or visit our Web site (www.milkweed.org).

My Green Manifesto:
Down the Charles River in Pursuit
of a New Environmentalism
By David Gessner

The Nature of College:
How a New Understanding of Campus Life
Can Change the World
By James J. Farrell

The Future of Nature:
Writing on a Human Ecology from Orion *Magazine*
Edited and introduced by Barry Lopez

Hope, Human and Wild:
True Stories of Living Lightly on the Earth
By Bill McKibben

MILKWEED EDITIONS

Founded as a nonprofit organization in 1980, Milkweed
Editions is an independent publisher. Our mission is to
identify, nurture and publish transformative literature, and
build an engaged community around it.

JOIN US

In addition to revenue generated by the sales of books we
publish, Milkweed Editions depends on the generosity of
institutions and individuals like you. In an increasingly
consolidated and bottom-line-driven publishing world, your
support allows us to select and publish books on the basis
of their literary quality and transformative potential. Please
visit our Web site (www.milkweed.org) or contact us at
(800) 520-6455 to learn more.

Milkweed Editions, a nonprofit publisher, gratefully acknowledges sustaining support from Amazon.com; Emilie and Henry Buchwald; the Bush Foundation; the Patrick and Aimee Butler Foundation; Timothy and Tara Clark; the Dougherty Family Foundation; Friesens; the General Mills Foundation; John and Joanne Gordon; Ellen Grace; William and Jeanne Grandy; the Jerome Foundation; the Lerner Foundation; Sanders and Tasha Marvin; the McKnight Foundation; Mid-Continent Engineering; the Minnesota State Arts Board, through an appropriation by the Minnesota State Legislature and a grant from the National Endowment for the Arts; Kelly Morrison and John Willoughby; the National Endowment for the Arts; the Navarre Corporation; Ann and Doug Ness; Jörg and Angie Pierach; the Carl and Eloise Pohlad Family Foundation; the RBC Foundation USA; the Target Foundation; the Travelers Foundation; Moira and John Turner; and Edward and Jenny Wahl.

Interior design by Connie Kuhnz
Typeset in Minion Pro
by BookMobile Design and Publishing Services
Printed on acid-free 50% postconsumer waste paper
by Malloy Incorporated